REMOTE SENSING OF ATMOSPHERE AND OCEAN FROM SPACE:
MODELS, INSTRUMENTS AND TECHNIQUES

ADVANCES IN GLOBAL CHANGE RESEARCH

VOLUME 13

Editor-in-Chief

Martin Beniston, *Institute of Geography, University of Fribourg, Perolles, Switzerland*

Editorial Advisory Board

B. Allen-Diaz, *Department ESPM-Ecosystem Sciences, University of California, Berkeley, CA, U.S.A.*
R.S. Bradley, *Department of Geosciences, University of Massachusetts, Amherst, MA, U.S.A.*
W. Cramer, *Department of Global Change and Natural Systems, Potsdam Institute for Climate Impact Research, Potsdam, Germany.*
H.F. Diaz, *Climate Diagnostics Center, Oceanic and Atmospheric Research, NOAA, Boulder, CO, U.S.A.*
S. Erkman, *Institute for Communication and Analysis of Science and Technology – ICAST, Geneva, Switzerland.*
M. Lal, *Centre for Atmospheric Sciences, Indian Institute of Technology, New Delhi, India.*
U. Luterbacher, *The Graduate Institute of International Studies, University of Geneva, Geneva, Switzerland.*
I. Noble, *CRC for Greenhouse Accounting and Research School of Biological Sciences, Australian National University, Canberra, Australia.*
L. Tessier, *Institut Mediterranéen d'Ecologie et Paléoécologie, Marseille, France.*
F. Toth, *International Institute for Applied Systems Analysis, Laxenburg, Austria.*
M.M. Verstraete, *Space Applications Institute, EC Joint Research Centre, Ispra (VA), Italy.*

The titles published in this series are listed at the end of this volume.

REMOTE SENSING OF ATMOSPHERE AND OCEAN FROM SPACE: MODELS, INSTRUMENTS AND TECHNIQUES

Edited by

Frank S. Marzano

and

Guido Visconti

*Center of Excellence CETEMPS,
University of L'Aquila, L'Aquila, Italy*

KLUWER ACADEMIC PUBLISHERS
DORDRECHT / BOSTON / LONDON

A C.I.P. Catalogue record for this book is available from the Library of Congress.

ISBN 1-4020-0943-7

Published by Kluwer Academic Publishers,
P.O. Box 17, 3300 AA Dordrecht, The Netherlands.

Sold and distributed in North, Central and South America
by Kluwer Academic Publishers,
101 Philip Drive, Norwell, MA 02061, U.S.A.

In all other countries, sold and distributed
by Kluwer Academic Publishers,
P.O. Box 322, 3300 AH Dordrecht, The Netherlands.

Printed on acid-free paper

All Rights Reserved
© 2002 Kluwer Academic Publishers and copyright holders
as specified on appropriate pages within.
No part of this work may be reproduced, stored in a retrieval system, or transmitted
in any form or by any means, electronic, mechanical, photocopying, microfilming, recording
or otherwise, without written permission from the Publisher, with the exception
of any material supplied specifically for the purpose of being entered
and executed on a computer system, for exclusive use by the purchaser of the work.

Printed in the Netherlands.

Contents

Contributors VII

Acknowledgements IX

CHAPTER 1: MODELS AND INVERSION METHODS FOR REMOTE SENSING 1

Atmospheric Backscatter and Lidars 3
   G.P. Gobbi and F. Barnaba

Precipitation Modeling for Inversion Purposes.. 19
   P. Bauer

Profile Retrieval Estimation Techniques... 35
   E. R.. Westwater

Bayesian Techniques in Remote Sensing.. 49
   N. Pierdicca

CHAPTER 2: ATMOSPHERIC REMOTE SENSING BY MICROWAVE AND VISIBLE-INFRARED SENSORS 65

Atmospheric Temperature Analysis using Microwave Radiometry 67
   S. English

Microwave Limb Sounding... 79
   S. Buehler

Retrieval of Integrated Water Vapour and Cloud Liquid Water Contents 89
L. Eymard

Precipitation Retrieval from Spaceborne Microwave Radiometers and Combined Sensors 107
F.S. Marzano, A. Mugnai and J.F. Turk

Clouds and Rainfall by Visible-Infrared Radiometry 127
V. Levizzani

Tropospheric Aerosols by Passive Radiometry 145
G.L. Liberti and F. Chéury

CHAPTER 3: OCEAN REMOTE SENSING BY MICROWAVE AND VISIBLE-INFRARED SENSORS 163

Sea Modeling by Microwave Altimetry 165
S. Pierini

Sea Scatterometry 183
M. Migliaccio

Sea Surface Characterization by Combined Data 201
L. Santoleri, B.B. Nardelli and V. Banzon

CHAPTER 4: DATA ASSIMILATION AND REMOTE SENSING PROGRAMS 215

Mesoscale Modeling and Methods Applications 217
R. Ferretti

COST Action 712: Microwave Radiometry 231
C. Matzler

Contributors

FRANCESCA BARNABA - Istituto di Scienze dell'Atmosfera e del Clima (ISAC) Sezione di Roma, Consiglio Nazionale delle Ricerche (CNR) - Via del Fosso del Cavaliere 100, 00133 Roma (Italy). E-mail: f.barnaba@isac.cnr.it

PETER BAUER - Research Department, Satellite Section, European Center for Medium range Weather Forecast (ECMWF) - Shinfield Park, Reading, Berkshire RG2 9AX, UK. E-mail: peter.bauer@ecmwf.int

STEFAN BUEHELER - Institute of Environmental Physics, University of Bremen - FB 1 p.o. box 330440; 28334 Bremen (Germany). email: sbuehler@uni-bremen.de

BRUNO BUONGIORNO NARDELLI - Istituto di Scienze dell'Atmosfera e del Clima (ISAC) Sezione di Roma, Consiglio Nazionale delle Ricerche (CNR) - Via del Fosso del Cavaliere 100, 00133 Roma (Italy). E-mail: b.buongiorno@isac.cnr.it

FREDERIC CHEURY - Laboratoire de Meterologie Dinamique, Conseil Nationale Recerches Scientifique (LMD-CNRS) - Paris (France), f.cheruy@lmd.jussieu.fr

STEPHEN J. ENGLISH - Satellite Radiance Assimilation Group, UK Met Office - London Road Bracknell Berkshire RG12 2SZ (United Kingdom). E-mail: stephen.english@metoffice.com

LAURENCE EYMARD - CETP/IPSL/CNRS, Université St Quentin-Versailles - Avenue de l'Europe 10-12, 78140 Vélizy (France). E-mail: Laurence.Eymard@cetp.ipsl.fr

ROSSELLA FERRETTI - Dipartimento di Fisica, Università dell'Aquila – Via Vetoio, 67010 Coppito (L'Aquila, Italy). E-Mail: rossella.ferretti@aquila.infn.it

GIAN PAOLO GOBBI - Istituto di Scienze dell'Atmosfera e del Clima (ISAC) Sezione di Roma, Consiglio Nazionale delle Ricerche (CNR) -

Via del Fosso del Cavaliere 100, 00133 Roma (Italy). E-mail: g.gobbi@isac.cnr.it

GIAN LUIGI LIBERTI - Istituto di Scienze dell'Atmosfera e del Clima (ISAC) Sezione di Roma, Consiglio Nazionale delle Ricerche (CNR) - Via del Fosso del Cavaliere 100, 00133 Roma (Italy). E-mail: g.liberti@isac.cnr.it

FRANK S. MARZANO - Dipartimento di Ingegneria Elettrica, Centro di Eccellenza CETEMPS, Università dell'Aquila - Monteluco di Roio, 67040, L'Aquila (Italy). E-Mail: marzano@ing.univaq.it

CHRISTIAN MATZLER - Institute of Applied Physics, University of Bern - Sidlerstr. 5, CH-3012 Bern (Switzerland) – E-mail: matzler@iap.unibe.ch,

MAURIZIO MIGLIACCIO - Onde Elettromagnetiche, Università degli Studi di Napoli Parthenope - Via Acton 38, 80133 Napoli (Italy). E-mail: maurizio.migliaccio@uninav.it

ALBERTO MUGNAI - Istituto di Scienze dell'Atmosfera e del Clima (ISAC) Sezione di Roma, Consiglio Nazionale delle Ricerche (CNR) - Via del Fosso del Cavaliere 100, 00133 Roma (Italy). E-mail: a.mugnai@isac.cnr.it

NAZZARENO PIERDICCA - Dipartimento di Ingegneria Elettronica, Università "La Sapienza" di Roma – Via Eudossiana 18, 00184 Roma (Italy). E-Mail: pierdicca@die.ing.uniroma1.it

STEFANO PIERINI - Dipartimento di Fisica, Università dell'Aquila – Via Vetoio, 67010 Coppito (L'Aquila, Italy). E-Mail: stefano.pierini@aquila.infn.it

ROSALIA SANTOLERI - Istituto di Scienze dell'Atmosfera e del Clima (ISAC) Sezione di Roma, Consiglio Nazionale delle Ricerche (CNR) - Via del Fosso del Cavaliere 100, 00133 Roma (Italy). E-mail: l.santoleri@isac.cnr.it

JOSEPH F. TURK - Naval Research Laboratory, Marine Meteorology Division - 7 Grace Hopper Avenue, Monterey, CA 93943-5502 USA. E-mail: turk@nrlmry.navy.mil

VINCENZO LEVIZZANI - Istituto di Scienze dell'Atmosfera e del Clima (ISAC) Sezione di Bologna, Consiglio Nazionale delle Ricerche (CNR) - via Gobetti 101, 40129 Bologna (Italy). E-mail: v.levizzani@isac.cnr.it

ED R. WESTWATER - Cooperative Institute for Research in Environmental Sciences, University of Colorado/NOAA, Environmental Technology Laboratory - 325 Broadway MS R/E/ET1 80305 Boulder (CO, USA). Email: Ed.R.Westwater@noaa.gov

Acknowledgements

The idea to held the first International Summer School on Atmospheric and Oceanic Sciences (ISSAOS) in L'Aquila was a sort of *coupe de foudre* during a Spring warm day. Nothing romantic, of course, but simply timely: during an unexpected discussion we realized, on one hand, the emergence of a larger and larger remote sensing activity at the University of L'Aquila and, on the other hand, the national academic interest to start a cycle of International Summer Schools on atmospheric and oceanic physics. The basic ingredients were soon identified: a topic of general scientific interests, a high quality selection of lecturers and, last but not least, a pleasant venue.

ISSAOS-2000 was in the period September 4-8 at the College of Engineering of the University of L'Aquila, amazingly placed on a hill at about 1000 m. The city of L'Aquila offered the historical landscape of five days of full immersion. More than 70 students responded to the call and agreed to follow more than 35 hours of lectures. The conclusive social dinner, amused by spontaneous music, was the deserved prize for all participants and organizers.

The primary audience of ISSAOS-2000 was constituted by graduate students in remote sensing and geosciences, Ph.D. or equivalent students, remote sensing researchers specialist in one single field, willing to have a basic understanding of spaceborne remote sensing capabilities for atmospheric and ocean remote sensing with current established techniques and a research perspective.

This book is essentially the collection of the lectures, given during ISSAOS-2000 by the lecturers coming both from Europe and USA. Its goal is to provide a broad panorama of spaceborne remote sensing techniques, at both microwave and visible-infrared bands and by both active and passive sensors, for the retrieval of atmospheric and oceanic parameters. A significant emphasis is given to the physical modelling background, instrument potential and limitations, inversion methods and applications. Topics on international remote sensing programs and assimilation techniques into numerical weather forecast models are also touched in the last chapter.

Since the main purpose of this book is to offer a well selected introductory material, each contribution is maintained within a limited number of pages, giving emphasis to the basic concepts and to an appropriate and abundant list of references. Even though this book is published after almost two years with respect to the end of ISSAOS-2000, we believe it represents a valuable sample of the state-of-the art on remote sensing of the atmosphere and ocean. In this regard, on January 2002 we asked the contributors to include up-to-date concepts and references within their sections.

As mentioned before, the organization of an International Summer School is a complex task, especially at its first edition. We acknowledge the financial contribution to ISSAOS-2000 coming from: the Italian Space Agency (ASI), the Science and Technology Park of Abruzzo (PSTd'A), the IEEE Geoscience and Remote Sensing Society – Central and South Italy Chapter, the Department of Electrical Engineering and the Department of Physics of the University of L'Aquila, the Italian Remote Sensing Association (AIT) and the Italian Inter-University Centre for Atmosphere and Hydrophere Physics (CIRFAI). We are also grateful to the Dean of the College of Engineering of the University of L'Aquila for the use of rooms and facilities.

The first personal acknowledgment goes to Simona Zoffoli of the Italian Space Agency (ASI) who enthusiastically supported the ISSAOS-2000 idea. Errico Picciotti, Tonia Fortunato, Adelaide Memmo and Fabrizio Mancini were irreplaceable for their support and continuous availability before and during ISSAOS-2000. A further acknowledgment is due to Ortensia Ferella, Pierlugi Strinella and Daniele Masoni of the International Science Services who have made available all their professionalism for the organization of the meeting. A very dedicated and warm acknowledgment goes to Simona Marinangeli, responsible for the editorial format of this manuscript: her effective and efficient work has made this hopefully important publication possible.

L'Aquila – April 3, 2002

Frank S. Marzano and Guido Visconti

Chapter 1

Models and Inversion Methods for Remote Sensing

Atmospheric Backscatter and Lidars

G. P. GOBBI AND F. BARNABA
CNR, Istituto di Scienze dell'Atmosfera e del Clima, Rome, Italy

Key words: Lidar, Aerosols, Climate

Abstract: This paper will provide an introduction to the laser radar (lidar) technique. We shall discuss the basic components of lidars and why these remote-sensing systems are very efficient at detecting atmospheric minor constituents with high spatial and temporal resolution. In particular, we shall address lidar observations of atmospheric aerosols, one of the major unknowns in the Earth's climate system. The fundamental tools for a quantitative retrieval of tropospheric aerosol backscatter and extinction on the basis of single-wavelength lidar observations will be given.

1. INTRODUCTION

LIDAR is an acronym for LIght Detection And Ranging. By exploiting the same principle of radars, these active remote-sensing systems operate by sending laser pulses into the atmosphere and recording the backscattered light by means of telescopes provided with highly sensitive photodetectors. Analysis of the signal intensity versus return time then allows for range-resolved definition of the physical and chemical properties of the scattering media. Since the development of the first lasers in the sixties, lidars have been employed in the remote observation of atmospheric trace gases, aerosols and clouds (e.g., Reagan et al., 1989). In fact, thanks to their high vertical and temporal resolution lidars provide an ideal tool to retrieve information about these constituents in regions of the atmosphere usually difficult to investigate.

Currently, a growing number of lidars is employed in the observation of atmospheric aerosols and clouds. The scientific interest in these studies resides in the role these particles play in controlling the Earth's climate. In fact, aerosol particles both scatter and absorb solar radiation. Furthermore, they interact with clouds by providing cloud condensation nuclei and by altering the droplet number concentration, with strong impact on both solar radiation and the hydrological cycle (e.g., Ramanathan et al., 2001).

Aerosols model-simulated climatic effects tend to counterbalance the expected warming due to the increase in greenhouse gases (e.g., Charlson, 1995). However, some indetermination still exists on the sign of this forcing (warming or cooling) in the presence of absorbing particles as desert dust and black carbon (Andreae, 2001). In fact, radiative effects of aerosols, particularly in the case of absorbing material, strongly depend on their altitude (Liao and Seinfeld, 1998), and this parameter is virtually impossible to evaluate by means of passive techniques. This explains the interest in studying such kind of particles by lidar.

In this chapter, after providing an introduction to the lidar technique, we shall focus on methods employed to study atmospheric aerosols and thin clouds. Reference to scientific journal bibliography will be essential. If interested in a more formal and/or thorough description of lidar techniques the reader is referred to the books of Measures (1984) and Hinkley (1976). A good overview addressing the role of atmospheric aerosols in climate can be found in Charlson (1995), with updating in Haywood and Boucher (2000) and in Ramanathan et al. (2001). Finally, a description of most of the lidar systems operating in the world, together with related literature and operators can be found at the lidar directory web site kept by NASA Langley (reported in the reference section).

2. THE LIDAR TECHNIQUE

Lidars commonly employ a laser source to emit a series of monochromatic, collimated light pulses which are sent towards the target to be studied. The scattered light is collected by optical telescopes, bandpass filtered to increase the signal to noise ratio, and then revealed by photo-detectors. The resulting electric signal is finally digitized to be stored and analyzed. When emitter and receiver are co-located we talk about monostatic lidars, conversely, we deal with bistatic lidars. In this work we will only address monostatic systems.

Even if based on the same concept, several lidar techniques have been developed to study different atmospheric parameters. In fact, various types of scattering/absorption processes can be exploited to retrieve information on the atmospheric status:

1. The frequency shift of light due to the Doppler effect is employed to derive wind speed (Doppler Lidar);
2. Spectral absorption of light by air molecules is the basis for the Differential Absorption Lidar (DIAL) that employs an on-line off-line comparison to retrieve concentration profiles of atmospheric gases;
3. Raman lidars measure the wavelength-shifted return from selected molecules produced by the inelastic (Raman) scattering to derive mixing ratio profiles of such molecules (e.g. water vapor);
4. Elastic scattering from molecules and aerosol is the physical process at the base of aerosol and Rayleigh (measuring molecular density and temperature) lidars.

Set up of the most frequently encountered lidar systems is based on the following components:

Emitter: Pulsed, solid state lasers are usually employed. Wavelengths range from the ultra-violet (UV) to the near infrared (IR). Most commonly employed lasers are Nd:YAG ($\lambda=1064$ nm, often operated with 2nd (532 nm) and/or 3rd (355 nm) harmonic generators), Ruby ($\lambda=694$ nm), Alexandrite (tunable in the range $\lambda=701-818$ nm), Titanium:Sapphire ($\lambda=720-960$ nm) and diode ($\lambda=450-32000$ nm) ones. Pulse energy and repetition rate can range from the μJ at several kHz, to 1-10 J in the 1-10 Hz range;

Receiver: Standard optical telescopes are employed to collect the returning light. Lidar telescopes range from single lens ones, with surface area as small as 0.01 m^2, to multiple-mirror ones, with surface area of the order of 1-10 m^2;

Bandpass: Background noise is commonly reduced by means of 1) interference filters (typical bandwidths $\Delta\lambda=0.1-10$ nm), 2) monocromators, 3) interferometers;

Detector: Most frequently photomultipliers (PM) are adopted to detect the signal. However, avalanche photo diodes (APD) are also well established. Photomultipliers are more effective in the UV-visible range, while photodiodes are more suited for visible-IR use. Typical gain needed for lidar detectors spans the range 10^5-10^7;

Signal acquisition: The electric signal generated by the detector is either converted by A/D devices or measured by means of photon counting techniques. The first approach is adopted in the presence of intense signals (usually close range returns), while the second well applies to faint signals. 10-100 MHz sampling rates or counting rate capability are required, respectively.

Laser beams have typical divergence of the order of 1 mrad. This converts into a spot of 1 m diameter at 1 km and of 10 m at 10 km. Considering a typical lidar range resolution of 10-100 m, the resulting sampling volume is rather small. Laser repetition rates in the range of 10-100

Hz, coupled with pulse energy of the order of 100 mJ allow to obtain good aerosol profiles by averaging over times of the order of one minute. These characteristics permit to achieve a detailed description of atmospheric constituents and processes ranging from the micro to the macro scale.

Instrumental set up and analysis schemes are equally important for the success of lidar observations. Both these aspects need to be carefully addressed to obtain quantitative information from the detected signal. The attention will be devoted here to the retrieval of geophysical information from a monostatic lidar trace generated by elastic backscatter (scattering angle $\theta=180°$).

3. ATMOSPHERIC SCATTERING

Upon travelling trough the atmosphere, electromagnetic (e.m.) radiation is scattered and absorbed by both molecules and particles. Elastic scattering is rather efficient when scatterer size is comparable to the impinging wavelength λ. For this reason, backscatter from the less abundant aerosols is comparable to the one generated by the much more abundant molecules.

A key parameter of a scattering process is the scattering cross section, σ_{sca}. If I_0 is the intensity of the incident radiation, the energy removed by the scattering process, E_{sca}, is given by $E_{sca} = I_0 \cdot \sigma_{sca}$. The scattering cross section is linked to the geometric cross section of the scatterer ($G=\pi r^2$) via the scattering efficiency parameter, $Q_{sca} = \sigma_{sca} / G$.

The Rayleigh scattering theory was developed to describe the interaction between solar radiation and air molecules (e.g. Kerker (1969)). In fact, it can be applied when the dimension of the scatterer is much smaller than the wavelength of the impinging radiation. According to the Rayleigh theory, molecular scattering cross section is proportional to λ^{-4}, making air molecules much more efficient scatterers at blue, rather than red wavelengths (this explains the blue color of the sky).

When particle dimensions become comparable to the wavelength of incident light, the Mie scattering theory is the most common tool employed to study the radiation-particle interaction (e.g. Kerker (1969)). In this case, due to the variability of particles size distribution and refractive index, a general wavelength dependence of the scattering cross section cannot be determined (roughly, σ_{sca} is proportional to λ^{-1}). The Mie theory describes the scattering of spherical particles with size comparable to the incident wavelength, this means that it applies well to liquid aerosols, cloud and fog droplets. However, it leads to some degree of error in the case of solid non-spherical aerosols (e.g., dust, smoke and frozen particles).

Typical laser wavelengths of the order of one micrometer make lidars ideal instruments to remotely observe aerosols and thin clouds, which are made of particles of size ranging between hundredths and tens of microns. In this case, the term "Mie lidar" is often employed.

3.1 Useful formulas about cross sections

The differential backscatter cross section of an "average" air molecule at wavelength $\lambda(\mu m)$, and below an altitude $z=100$ km, can be computed as in Hinkley (1976):

$$d\sigma_m(\lambda, \pi) / d\Omega = 5.45 \, [\lambda(\mu m) / 0.55]^{-4} \times 10^{-28} \quad (cm^2 \, sr^{-1}) \quad [1]$$

(notice the characteristic λ^{-4} dependence of the Rayleigh scattering). The backscatter coefficient β_m at an altitude z is defined as the differential backscatter cross section per unit volume:

$$\beta_m(\lambda, z) = N_m(z) \cdot d\sigma_m(\lambda, \pi) / d\Omega \quad [2]$$

At sea level (z=0), with typical molecular density $N_m = 2.5 \times 10^{19}$ cm^{-3}, this leads to a backscatter coefficient (Hinkley, 1976):

$$\beta_m(\lambda, 0) = [\lambda(\mu m) / 0.55]^{-4} \times 1.39 \times 10^{-8} \quad (cm^{-1} \, sr^{-1}) \quad [3]$$

From the Rayleigh scattering theory, we find the extinction/backscatter ratio of air molecules to be a fixed value:

$$\sigma_m / \beta_m = 4 \pi / 1.5 = 8.378 \quad (sr) \quad [4]$$

Backscatter coefficients are expressed in $l^{-3} \cdot l^2 \cdot sr^{-1}$ units, i.e., $l^{-1} \cdot sr^{-1}$, with l standing for length unit. Conversely, extinction coefficients are expressed in l^{-1} units.

In the case of Mie scattering, the backscatter cross section $\sigma_B(r, \lambda, m)$ can be written as:

$$\sigma_B(r, \lambda, m) = \pi r^2 \, Q_B(x, m) \quad [5]$$

where r is the particle's radius and $Q_B(x, m)$ the backscatter efficiency, provided by the Mie theory. $Q_B(x, m)$ is a complex function of particles complex refractive index $m = n - ik$ and of the Mie parameter:

$$x = 2 \pi r / \lambda \quad [6]$$

In a similar fashion, the Mie extinction cross section σ_E (r, λ, m) can be written as:

$$\sigma_E(r, \lambda, m) = \pi r^2 Q_E(x, m) \quad [7]$$

where Q_E (x, m) is the extinction efficiency. Q_B and Q_E are expressed by an infinite series of Riccati-Bessel functions of x and m·x. To compute backscatter or extinction coefficients of a distribution of particles N(r) (cm^{-3}) the product σ_{Mie}·N(r) must be integrated over r. Therefore, aerosol backscatter is computed as:

$$\beta_a = \int_0^\infty Q_B \pi r^2 N(r) dr \quad [8]$$

A frequently used analytical formulation to represent aerosol size distributions is the log-normal one:

$$N(r) = \frac{dn}{d\log r} = \frac{N}{\sqrt{2\pi} \cdot \log \sigma} \exp\left[-\frac{(\log r - \log r_m)^2}{2\log^2 \sigma}\right] \quad [9]$$

where N is the particle concentration, r_m and σ the distribution modal radius and width, respectively. The log-normal function is widely employed because it reproduces quite well the bell-shaped dispersion (in log coordinates) characterizing natural aerosols.

4. THE LIDAR EQUATION

The lidar equation describes the amount of laser energy returning to the lidar detector from the distance R during the sampling time interval τ_d:

$$E(\lambda, R) = E_L \cdot (c\tau_d/2) \cdot (A_0/R^2) \cdot \xi(\lambda) \cdot \beta(\lambda, R) \cdot T^2(\lambda, R) \quad [10]$$

Where the symbols represent, respectively:
E_L : Laser pulse energy;
λ : laser wavelength;
R : Range;
A_0 : Area of lidar receiver (telescope);
$\xi(\lambda)$: Receiver spectral transmission at the wavelength λ;
$\beta(\lambda, R)$: Atmospheric backscatter coefficient at distance R and wavelength λ;

T (λ, R): Atmospheric transmittance between the lidar and the backscattering volume at distance R, at wavelength λ.

Similarly to radars, the return time of the backscattered laser pulse (dt) indicates the distance (R) of the scattering object: $R = c\,dt/2$, where c is the speed of light. Therefore, a 10 µs return time corresponds to $R = 1,500$ m, while 100 µs convert into $R = 15$ km. Conversely, the sampling time interval τ_d defines the lidar profile spatial resolution: $dR = c\,\tau_d / 2$.

The two variables bearing the geophysical information we want to retrieve by solving the lidar equation are: β (λ, R), the backscatter coefficient, and T (λ, R), the atmospheric transmittance. As previously introduced, the backscatter coefficient of a species y can be defined as:

$$\beta_y (\lambda, R) = N_y (R) \cdot d\sigma_{sy} (\lambda, \pi) / d\Omega \qquad [11]$$

where N_y and $d\sigma_{sy} (\lambda, \theta)/d\Omega$ represent the volume number concentration and the differential scattering cross section (at $\theta = 180°$), respectively. In the case of elastic backscatter from the atmosphere, β can be expressed as the sum of a molecular plus a particle contribution:

$$\beta_{m+a} (\lambda, R) = \beta_m (\lambda, R) + \beta_a (\lambda, R) \qquad [12]$$

where the subscripts "m" and "a" stand for the molecular and the aerosol (particulate) component, respectively. In fact, an aerosol-free atmosphere is only found above an altitude of 30-40 km. Upward these altitudes molecular, i.e., Rayleigh scattering dominates.

The atmospheric transmittance between the lidar site (R=0) and the point at distance R can be expressed as:

$$T(\lambda, R) = e^{-\int_0^R \sigma (\lambda, R) dR} \qquad [13]$$

Where $\sigma(\lambda, R)$ is the atmospheric extinction coefficient:

$$\sigma_y (\lambda, R) = N_y(R)\, \sigma_{sy} (\lambda) \qquad [14]$$

with $\sigma_{sy} (\lambda)$ the species extinction cross section. Again, in an atmosphere composed of molecules and aerosols we can express σ as:

$$\sigma_{m+a} (\lambda, R) = \sigma_m (\lambda, R) + \sigma_a (\lambda, R) \qquad [15]$$

4.1 Solving the lidar equation

By omitting the range and wavelength-dependence symbols and assuming the receiver spectral transmittance to be $\xi(\lambda)=1$ we can rewrite the lidar equation as:

$$E = E_L \cdot \frac{c\tau_d}{2} \cdot \frac{A_0}{R^2} \cdot (\beta_m + \beta_a) \cdot e^{-2\int_0^R (\sigma_m + \sigma_a)dR} \qquad [16]$$

Eq. (16) is underdetermined since it contains the two unknowns β_a and σ_a (β_m and σ_m are commonly obtained from eqs. (2) and (4) by employing either radio-soundings or atmospheric density models). Conversely, Eq. (16) needs some relationship linking the aerosol backscatter and extinction coefficients to be solved for either β_a or σ_a. Collis and Russell (1976) indicated as a general parameterization the following one:

$$\beta_a = \text{cost } \sigma_a^g \qquad [17]$$

with g depending on the specific properties of the scattering medium and generally spanning the range $0.67 \leq g \leq 1$. Several versions of this relationship have been used in the literature (e.g., Klett, 1985; Kovalev, 1993). However, the most frequent assumption in lidar inversion schemes is to set g=1, i.e., employ a constant ratio between backscatter and extinction coefficients. This is in spite of the fact that Klett (1985) showed that a good retrieval of backscatter in turbid atmospheres cannot be obtained when a constant extinction/backscatter ratio is employed. We shall discuss later some methods employed to determine the relationship between β and σ for tropospheric aerosols.

One of the most popular solutions of the lidar equation is the analytical one described by Klett (1985):

$$\beta_a(R) + \beta_m(R) = \frac{\exp(S'-S_c')}{\left[\dfrac{1}{\beta_c} + 2\int_R^{R_c} \dfrac{\exp(S'-S_c')dR}{B_a}\right]} \qquad [18]$$

Where $B_a = \beta_a/\sigma_a$. This solution has been found by defining the range corrected signal:

$$S(R) = \ln\left[E(R) \cdot R^2/\tau_d\right] \qquad [19]$$

which converts the lidar equation into:

$$\frac{dS}{dR} = \frac{1}{\beta_{m+a}} \frac{d\beta_{m+a}}{dR} - 2\sigma_{m+a} = \frac{1}{\beta_{m+a}} \frac{d\beta_{m+a}}{dR} - \frac{2\beta_{m+a}}{B_a} + 2\beta_m \left(B_a^{-1} - B_m^{-1}\right) \quad [20]$$

By defining now the new signal variable S':

$$S' - S'_c = S - S_c + \frac{2}{B_m} \int_R^{R_c} \beta_m dR' - 2 \int_R^{R_c} \frac{\beta_m}{B_a} dR' \quad [21]$$

Eq. (20) becomes:

$$\frac{dS'}{dR} = \frac{1}{\beta_{m+a}} \frac{d\beta_{m+a}}{dR} - \frac{2\beta_{m+a}}{B_a} \quad [22]$$

which is solved by the Klett equation after definition of the boundary values at the calibration range $R=R_c$.

The Klett solution of the lidar equation is a stable one. However, it contains the relationship $B_a = \beta_a/\sigma_a$. If B_a is assumed to vary as a function of aerosol properties, for instance $B_a = f(\sigma_a) = \beta_a/\sigma_a$ then an iterative process is necessary to retrieve the correct value of β_a (as previously mentioned, β_m is derived either from air density measurements or models).

5. BACKSCATTER AND DEPOLARIZATION RATIO

A typical lidar signal profile S_0 (after subtracting background noise) described by Eq. (16) is plotted in Figure 1a (curve 1). This measurement was taken in the presence of mineral dust aerosols between 0 and 6 km and of a cirrus cloud between 10 and 12 km. Also in Figure 1a are plotted the range–corrected signal ($S_1 = S_0 \cdot R^2$, curve 2), and the range, extinction–corrected signal ($S_2 = S_0 \cdot R^2 \cdot T^2$, curve 3). The molecular signal (S_{mol}) plotted in Figure 1a was determined solving the lidar equation for a purely molecular atmosphere, on the basis of an atmospheric density model. On the basis of Eqs. (2) and (4), S_{mol} allows for calibration of the lidar measurement by normalizing the two curves in a region where aerosol backscatter is expected to be negligible (remember that calibration at an aerosol-free point R_c is also required in the Klett solution above). Here calibration was performed in the region 7-8 km.

A common way to represent aerosol lidar data (and avoid dealing with the rapid decrease in atmospheric backscatter signal S) is the backscatter ratio:

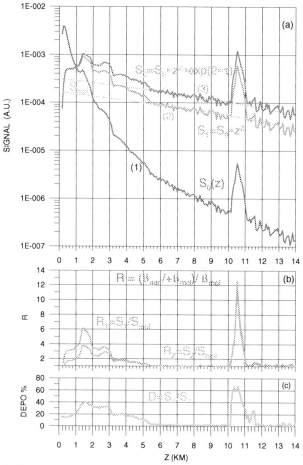

Figure 1. a) Signal S_0 observed by the 25-cm VELIS lidar at Rome on March 24, 2001 (curve 1). Curve 2 shows the range-corrected signal $S_1 = S_0 z^2$, while curve 3 depicts the range plus extinction-corrected signal $S_2 = S_0 z^2 e^{2\tau}$; b) backscatter ratio R_B for the two traces S_1 and S_2 in figure 1a; c) depolarization ratio for the measurement in figure 1a.

$$R_B = S_{tot}/S_{mol} = (\beta_a + \beta_m)/\beta_m \qquad [23]$$

that is the ratio of the total atmospheric backscatter to the molecular one. Since $\beta_a = \beta_m (R_B-1)$, its departure from $R_B=1$ indicates the presence of a contribution of aerosols to the total atmospheric backscatter.

In Figure 1b two plots of the backscatter ratio are presented: one before and one after correcting the lidar signal for the particle extinction, i.e., for curves 2 and 3 of Figure 1a, respectively. This representation clearly shows how strong can be the impact of extinction in the quantitative determination of aerosol backscatter and vice versa.

Laser light is often polarized. Polarization is an useful tool to investigate aerosol thermodynamic phase. In fact, by detecting the amount of depolarization, we can discriminate between liquids and solids, i.e., droplets and crystals. This is because spheres do not generate any depolarization when backscattering a polarized beam. Conversely, when hit by polarized light, non-spherical particles depolarize part of the light they backscatter. A common way to quantify depolarization of lidar signals is the linear depolarization ratio:

$$D = S_\perp / S_{//} \qquad [24]$$

where $S_\perp$ and $S_{//}$ represent the lidar signal detected on polarization planes perpendicular and parallel to the laser one, respectively. Therefore, polarization lidars need to operate by means of two detection channels.

Signal depolarization induced by air molecules does not exceed 0.4%. This is due to their non-symmetrical structure. However, if the band pass filtering is not narrow enough ($\Delta\lambda > 1$ nm) depolarization can increase due to the inclusion of some Raman scattering returns, e.g., Young (1980). Furthermore, non-ideal characteristics of the laser emission or of the lidar optics can also increase the degree of observed depolarization. In this respect, calibration of polarization lidars is always necessary.

A theoretical description of lidar depolarization caused by particles of size comparable to the wavelength is still incomplete. In fact, analytical solutions are obtained only for particles with an axis of symmetry, i.e., rather regular ones, e.g., Mishchenko et al. (1997). Actual observations show an upper leveling of the depolarization at values D≈40-50% in the case of mineral dust and at values D≈50-60% in the case of ice crystals. Such a behavior can be used to infer the aerosol/cloud phase and nature (e.g., Gobbi et al., 1998; Gobbi et al., 2000). The depolarization ratio of the lidar profile of Figure 1a is presented in Figure 1c. As mentioned, the return between 0 and 6 km is mainly due to Saharan dust, while between 10 and 12 km we have a cirrus cloud. Maximum depolarization is of the order of 40 and 60 %, respectively. Depolarization is an absolute measurement, therefore it is independent of extinction correction.

6. THE EXTINCTION/BACKSCATTER RATIO OF AEROSOLS

As mentioned in section 4.1, determining the range-dependent behavior of the aerosol extinction/backscatter ratio, $R_{eb}(R)$ is fundamental to quantify atmospheric extinction or backscatter coefficients from a lidar observation.

This goal is currently approached by two different methods: 1) by implementing complementary lidar channels to actually measure R_{eb}; 2) by employing aerosol scattering models to infer such a ratio.

The first method is implemented by either adding a Raman channel to measure the atmospheric molecular return alone (e. g., Ansmann et al., 1992) or by adopting a high spectral resolution technique (e.g., Piironen and Eloranta, 1994). In the first case, the only unknown in the Raman return (for instance from N_2) is the aerosol extinction (again, expected molecular backscatter is evaluated by a density measurement or a model atmosphere). This allows to retrieve the two unknowns (β_a and σ_a) by inverting the two independent lidar measurements. In this approach, differences existing between the Raman and the laser wavelengths can lead to different aerosol extinction, then introducing some degree of error in the retrieval. In the case of high spectral resolution, the lidar receiver must be capable of separating the different Doppler broadening of the laser line induced by fast moving molecules as opposed to the slower aerosols. Such a technique requires band pass filtering of the order of 1-2 pm, i.e., high technology components.

Both techniques (in particular the Raman one) have to deal with lower signal intensity compared to the full backscatter signal at the laser wavelength. This problem reduces the maximum range these systems can detect. Furthermore, both techniques involve more (and more sophisticated) detection channels, that is, an increase in the complexity and mass of the lidar system. In the case small portable systems are preferred, an aerosol model represents an easier way to provide the range-dependent R_{eb} to be employed in the inversion of a single wavelength lidar signal.

Hereafter we shall describe a model approach developed to find a direct functional relationship between extinction and backscatter for various kinds of aerosols. Once the aerosol nature is inferred, use of such a relationships allows for a recursive solution of the lidar equation, without need for other external information.

6.1 Studying R_{eb} in some typical atmospheric load conditions: Maritime, desert dust and continental aerosols

A numerical model to explore and parameterize tropospheric aerosol extinction/backscatter relationship has been developed following an approach already used to address stratospheric aerosols (Gobbi, 1995). In the troposphere, sea salt and desert dust particles represent the first and the second contributor to the global aerosol load, respectively (e.g., d'Almeida et al., 1991). In addition to these natural sources, a large fraction of the aerosol injected into the atmosphere has an anthropogenic origin. In particular, biomass and fossil fuel combustion as well as industrial activity can be held responsible for a large part of carbon containing species (organic and black carbon).

Model results collected here have been obtained to describe typical atmospheric conditions, such as maritime, desert dust and continental aerosol loads (e.g., Barnaba and Gobbi, 2001). The model approach is based on three main steps:

1) Study of the variability of the investigated aerosol size distribution and composition as observed in nature;

2) Computation of the extinction and backscatter properties of a large number of size distributions and composition derived from step 1);

3) Fit of the resulting data dispersion by means of analytical curves.

Step 1) is performed by finding in the scientific literature the variability range of the parameters describing the size distributions and the refractive index of specific aerosol types. Step 2) is performed by means of a Monte Carlo approach: 20,000 size distributions are generated for each aerosol type (maritime, desert dust, continental) and their extinction and backscatter coefficients (σ_a and β_a) computed by means of the Mie theory. Each distribution has parameters randomly generated within the variability boundaries determined at step (1). In step 3) the dispersion of the σ_a vs. β_a points obtained in step 2) is fitted by means of an analytical curve to provide the $\sigma_a = f(\beta_a)$ functional relationship and, implicitly, the $R_{eb} = f(\beta_a)$ relationship.

According to literature (e.g., Barnaba and Gobbi, 2001), three modes have been considered for the maritime aerosols: two modes characterizing the sea salt component (dry modal radii variability ranges being $r_{m1}=$ 0.05-0.1 μm and $r_{m2}=$ 0.4-0.6 μm, respectively) and one mode for the sulfate component ($r_{m3}=$0.02-0.1 μm). Two modes have been employed to characterize the desert dust ($r_{d1}=$0.02-0.08 μm and $r_{d2}=$0.3-1.5 μm). Three modes have been selected for the continental aerosol, these respectively describe the contribution of water soluble (ws), carbon containing (cc) and dust like (dl) particles ($r_{ws}=$0.021-0.077, $r_{cc}=$0.007-0.012, $r_{dl}=$0.43-0.5). In all cases each mode is described by one log-normal size distribution (see Eq. (9)). The variability range of the number concentration N, the width σ, the real and imaginary part of the refractive index of each mode have also been determined from the same sources.

Dispersion of the σ_a vs. β_a points for the continental, desert dust and maritime aerosol as obtained by the model are presented in Figure 2. In the case of desert dust, an additional scatter plot of σ_a vs. β_a has been generated by converting the relationships obtained by Mie scattering computations into relationships for randomly oriented spheroids (Figure 2b).

This has been achieved by employing the sphere versus spheroids results

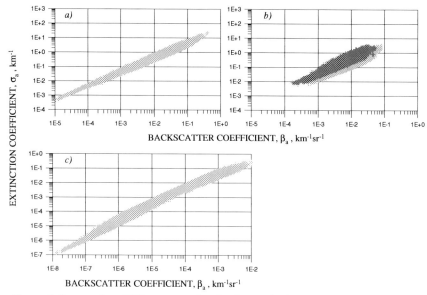

Figure 2. Scatter plot of 532 nm aerosol extinction vs. backscatter coefficients obtained for: a) continental, b) desert dust and c) maritime aerosols (20,000 points for each aerosol type). For desert dust, computations have been carried out for both spherical and non-spherical particles (light and dark gray crosses, respectively).

obtained by Mishchenko (1997) in simulating scattering from dust particles.

These results show that for the same backscatter coefficient, extinction of non-spherical particles can be much larger (over a factor of two) with respect to spherical ones.

Continuity found between the maritime aerosol scatter plot and both desert dust and continental aerosol ones allowed to define single curves (7-order polynomials in log-log coordinates) describing the ensembles of two aerosol types: maritime+desert dust and maritime+ continental. In the first case, two fitting curves have been determined: one describing the maritime + *spherical* dust model and the other for the maritime + *non-spherical* dust model.

Table 1. Coefficients of the 7th order polynomials (Eq.25) describing the extinction versus backscatter relationship for the ensemble of maritime+desert dust aerosols (M+D, spherical (s) and non-spherical (ns) results) and for maritime+continental aerosol (M+C).

Coeff.	a	b	c	d	e	f	g	h
M+D(s)	-6.26	-17.958	-18.043	-8.5650	-2.22173	-0.321168	-0.0243605	-7.5701E-4
M+D(ns)	-11.51	-26.245	-22.367	-9.4481	-2.23697	-0.300218	-0.0213845	-6.2928E-4
M+C	2.48	3.433	2.667	1.5084	0.47566	0.082961	7.41199E-3	2.6398E-4

By means of these results, inversion of the lidar signal in the presence of these aerosol types can be performed employing a single extinction/backscatter functional relationship of the kind:

$$\log_{10}(\sigma_a) = f(\beta_a) = a+bX+cX^2+dX^3+eX^4+fX^5+gX^6+hX^7 \quad [25]$$

where $X=\log_{10}(\beta_a)$. The model-derived coefficients of the polynomials for the three cases are reported in Table 1 (see also Barnaba and Gobbi, 2001).

7. CONCLUSIONS

An introduction to the lidar technique has been provided. We have seen how lidars represent an efficient tool to observe atmospheric minor constituents with high spatial and temporal resolution. In particular, we addressed atmospheric aerosols, one of the major unknowns in the Earth's climate system. The fundamental tools for a quantitative retrieval of aerosol extinction and backscatter on the basis of single-wavelength lidar observations have been presented. Even if lidar techniques allow to retrieve aerosol profiles with vertical resolution of a few meters and time resolution of the order of one minute, ground-based lidars only represent a point observation. In this respect, it can be easily foreseen that planned space-borne missions including both lidar and radiometric observations (as CALIPSO, expected to fly in 2003) constitute the natural evolution of the application of lidars to the definition of the global aerosol budget. Furthermore, our knowledge of atmospheric process will undergo an even larger leap when both aerosol and clouds will be observed from a space-borne platform by means of lidar and radar techniques as in the forthcoming ESA Earthcare mission. This is because aerosols and clouds constitute a unique ensemble, continually developing one into the other. This picture definitely indicates that lidar applications will soon conquer a large share of the global remote sensing activities.

8. REFERENCES

Andreae, M. O., The dark side of aerosol, Nature, 409, 671-672, 2001.

Ansmann, A., M. Riebesell, U. Wandinger, C. Weitkamp, E. Voss,, W. Lahmann and W. Michaelis, Combined Raman elastic-backscatter lidar for vertical profiling of moisture, aerosol extinction, backscatter and lidar ratio, Appl. Phys., B, 55, 18-28, 1992.

Barnaba, F, and G. P. Gobbi, Lidar estimation of tropospheric aerosol extinction, surface area and volume: maritime and desert dust cases, J. Geophys. Res., 106-D3, 3005-3018, 2001.

Charlson, R. J. (Ed.) Aerosol forcing of climate, J. Wiley, 416pp, New York, 1995.

d'Almeida, G., P Koepke, and E. P. Shettle, Atmospheric aerosols, 561 pp., A Deepak, Hampton, VA, 1991.

Gobbi, G.P., Lidar estimation of stratospheric aerosol properties: Surface, volume, and extinction to backscatter ratio, J. Geophys. Res., 100, 11,219-11,235, 1995.

Gobbi, G.P., G. Di Donfrancesco and A. Adriani, Stratospheric clouds physical properties during the Antarctic winter of 1995, J. Geophys. Res., 103, 10,859-10,874, 1998.

Gobbi, G. P., F. Barnaba, R. Giorgi and A. Santacasa, Altitude-resolved properties of a Saharan dust event over the Mediterranean, Atmos. Env., 34, 5119 - 5127, 2000.

Haywood, J., and O. Boucher, Estimates of the direct and indirect radiative forcing due to tropospheric aerosols: a Review, Rev. Geophys., 38, 4, 513-543, 2000.

Hinkley, E. D. (Ed.), Laser monitoring of the atmosphere, Springer Verlag, 380pp, New York, 1976.

Kerker, M., The scattering of light and other electromagnetic radiation, Acad. Press, New York, 1969

Klett, J. D., Lidar inversion with variable backscatter/extinction ratios, Applied Optics, 24, 1638-1643, 1985.

Kovalev, V. A., Lidar measurement of the vertical aerosol extinction profiles with range-dependent backscatter to extinction ratio, Applied Optics, 30, 6053-6065, 1993.

Liao, H., and J.H. Seinfeld, Radiative forcing by mineral dust aerosols: Sensitivity to key variables, J. Geophys. Res., 103, 31637-31645, 1998.

Measures, R. M., Laser Remote Sensing, J. Wiley, New York, 510 pp., 1984.

Mishchenko, M. I., L. D. Travis, R. A. Kahn, and R. A. West, Modeling phase functions for dustlike tropospheric aerosols using a shape mixture of randomly oriented polydisperse spheroids, J. Geophys. Res., 102, 16,831-16,847, 1997.

Piironen, P. and E. W. Eloranta, Demonstration of a high-spectral-resolution lidar based on a Iodine absorption filter, Opt. Lett., 19, 234-236, 1994.

Ramanathan, V., P.J. Crutzen, Kiehl, J.T., and Rosenfeld, D., Aerosols, Climate, and the Hydrological Cycle, Science, 294, 2119-2124, 2001.

Reagan, J. A., M. P. McCormick, and J. D. Spinhirne, Lidar sensing of aerosols and clouds in the troposphere and stratosphere, Proc. IEEE, 77, 433-447, 1989.

Young, A. T., Revised depolarization corrections for atmospheric extinction, Appl. Opt., 19, 3247-3248, 1980.

LIDAR DIRECTORY: http://www-rab.larc.nasa.gov/lidar/

Precipitation Modeling for Inversion Purposes

P. BAUER
Research Department, European Centre for Medium-Range Weather Forecasts, UK

Key words: Precipitation retrieval, microwave radiative transfer, satellite observations.

Abstract: Precipitation retrieval from space which makes use of physical forward radiative transfer modeling requires the proper treatment of all sources of radiation scattering and emission in atmosphere, clouds, and interaction with surfaces. This involves the evaluation of the significance of individual effects with respect to the observed signals. In case of microwave radiation, these effects are the three-dimensional distribution of temperature, humidity, and hydrometeor concentrations, particle size distributions, and particle composition and shape. On the technical side of the problem, the accuracy of the radiative transfer model and the simulation of the radiometer's imaging specifications are important. Most of the above effects have been described in the past thus a certain background for the generation of retrieval databases from radiative transfer simulations is available. However, there are major drawbacks at the current state of precipitation retrieval: (1) even though individual radiation processes are well described, no synthesis is available combining the best available models; (2) the errors of radiative transfer in realistic clouds are unknown thus limit its use in numerical prediction models together with satellite data; (3) the input to the radiative transfer models, i.e., cloud and precipitation models seem insufficient for the application of retrievals beyond regional studies.

1. INTRODUCTION

Principally, inversion techniques may be developed from simulations of satellite observations using cloud models ('physical') or by calibration of observations with available validation data ('empirical') (Smith et al., 1998). Figure 1 shows an approach of physical precipitation retrieval scheme development to be applied to passive and active microwave observations.

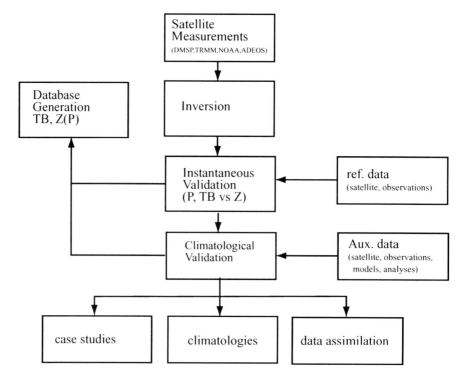

Figure 1: Logical flow of precipitation retrieval algorithm development. **TB** denotes the observable brightness temperature vector, **Z** is the radar reflectivity vector, and **P** denotes the atmospheric/surface parameter vector.

A retrieval database may be generated from radiative transfer simulations using an appropriate set of cloud model profiles and other input parameters which cover the variability of naturally occurring situations. From statistical (least-squares methods, neural networks) or probabilistic (Bayesian) inversion methods, the best estimate may be obtained and validated with independent reference data. In case of systematic and/or inacceptable errors, database and/or inversion method have to be updated. After the validation of instantaneous retrievals, climatological retrievals may follow depending on the purpose of the desired algorithm. These involve other problems such as sampling issues (not discussed here). Finally, algorithm application in cases studies or, more operationally, in data assimilation is possible. The flow diagram outlined in Figure 1 represents the basis for several operational and experimental algorithms (e.g. Kummerow et al., 1996; Haddad et al., 1998; Bauer et al., 2001).

The most crucial part of algorithm performance is the retrieval database which consists of cloud physics and radiative transfer elements. The following sections will address the most significant issues of both elements. Their ingredients are displayed with more detail in Figure 2.

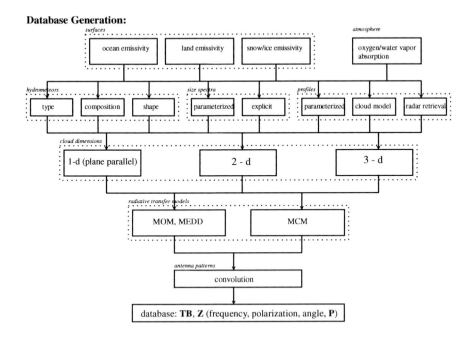

Figure 2: Problems relevant to microwave radiative transfer simulation in precipitation (MOM: matrix operator model, MEDD: modified Eddington model, MCM: Monte Carlo model).

2. SURFACE AND ATMOSPHERE

As illustrated in Figure 2, surfaces and cloud-free atmospheres represent the background for cloud radiative effects. The emissivity of water surfaces at frequencies between 1-200 GHz is fairly low thus the atmospheric contribution to the observed radiances (expressed as blackbody equivalent temperatures = brightness temperatures, TB) is strong. For this reason, retrieval of atmospheric moisture and temperature is comparably accurate over water surfaces and operationally implemented in weather forecast models (e.g. Eyre et al., 1993; Deblonde, 1999).

The modeling of water surface emissivity has to cover its dependence on frequency, temperature, salinity (ν<5 GHz), and roughness as a function of wind speed (and direction). The latter often involves an approach separating the effects of large gravity waves, capillary waves, and foam by use of wave models. The accuracy of these models is of the order of 1-5 K depending on frequency when comparing co-located satellite and near-surface observations (Eymard et al. 1999).

Land surfaces are much more difficult to describe because of the large number and complexity of contributing factors. To the first order, skin temperature, soil moisture and roughness, vegetation coverage and its water content are the free variables. However, models which may also include soil composition or the variability of vegetation are only available from some experiments thus their general applicability is questionable [see overview in Chanzy and Wigneron (1999)].

Atmospheric absorption is comparably well understood for line absorption (water vapor and oxygen) while continuum absorption is somewhat difficult to estimate from laboratory experiments. Most widely used are parameterizations which combine all effects integrating absorption contributions from all significant lines and the continuum as a function of temperature, pressure, and humidity. For example, the Millimeter Propagation Models (MPM's; e.g. Liebe et al., 1992) are commonly used for background absorption calculations.

3. HYDROMETEORS

3.1 Input data

Aiming at generalized retrievals (e.g. for global applications), the retrieval databases have to cover a large variability of cloud/precipitation cloud types and evolution stages. Principally, the required input data is not produced from a single data source. It thus depends on the application purpose and the degree of detail in radiative transfer which input data source to choose. Options are:
- Cloud profiles which are constructed on a single-column basis from radiosonde/global model profiles of temperature and humidity (Liu and Curry, 1992; Bauer and Schluessel, 1993).
- Output from numerical cloud simulation models providing two or three-dimensional cloud fields (Mugnai and Smith 1988).
- Data from ground-based or spaceborne radars which is converted to three-dimensional fields of precipitation in combination with the above (Viltard et al., 1999).

The single-column approach has the advantage of a better representativeness in terms of cloud type; however, the necessary simplification of radiative transfer to one dimension represents a serious drawback in view of real cloud geometries. Radar data will certainly provide a very accurate distribution of liquid precipitation patterns but radar signatures at frequencies below 35 GHz are not well modelled for frozen precipitation and negligible for cloud droplets. Therefore its use is only recommended in combination with other data sources filling in the missing constituents.

More recently, cloud model simulations began to dominate the input data generation because they allow the representation of consistent hydrometeor distributions in three dimensions over cloud evolution. Since these simulations usually originate from studies on tropical convection only very little data is available for shallow convection, warm rain, and frontal systems on a spatial resolution needed for the inclusion of sub-pixel variability (1-3 km) in the retrieval databases.

3.2 Hydrometeors

If the input data source covers several particle types, their optical properties have to be defined. The following list gives an overview of the important issues:

Cloud droplets: At millimeter wavelengths, absorption and scattering of cloud droplets mainly depends on cloud liquid water content, its temperature and the sounding frequency. Effects of particle size and drop size distribution (DSD) are negligible so that the assumption of monodisperse clouds is rather accurate. For temperatures below 270 K, the available models computing water permittivity give different answers (e.g. Lipton et al., 1999) which represents the only significant uncertainty for this hydrometeor type once liquid water concentrations are assumed to be correct.

Rain drops: Precipitating liquid water is a major research target thus a large number of studies in radar meteorology, satellite remote sensing, cloud and large-scale modeling are available. Particle sizes generally range from 0.2 mm to 10 mm thus both scattering and absorption of microwave radiation is important. Major uncertainties are the above mentioned water permittivity of supercooled drops, particle DSD's, and particle shapes. Since particle DSD's have the dominant effect on radar reflectivity, many radar studies deal with this topic (e.g. McKague et al. 1998). In retrieval databases, DSD's are mostly approximated by gamma-type functions adjusting shape parameter and offset (N_o) to local conditions. Recently, particle shape received some attention for passive microwave retrievals by Czekala and Simmer (1998); however, for downward looking radiometers the shape effect is rather small compared to other factors.

Snow, graupel, hail: Precipitating ice represents the biggest unknown in database generation. Cloud model parameterizations of ice microphysics are rather uncertain so that ice contents and the evolution of ice particle shape, density, and size distribution are very difficult to simulate. In most cases, constant densities (e.g. 0.9 g cm^{-3} for hail, 0.4 g cm^{-3} for graupel and 0.1 g cm^{-3} for snow) are assumed. Modifications to this may have different reasons: (1) Cloud models may provide more information on size distributions than the single-moment types (=water contents). For example, a double-moment representation (Ferrier, 1994; Swann, 1998) calculates water contents and total number concentrations over time, while a triple-moment representation may also calculate the particle volume of frozen species. (2) 'A posteriori' adjustments of particle spectra and size-dependent densities may be applied for improving the match between radiative transfer simulations and observations (Panegrossi et al., 1998; Schols et al., 1999). The latter, however, may be in conflict with the cloud model parameterizations but seems justified by the lack of better cloud model simulations.

The effects of particle shape on microwave emission/scattering are also difficult to generalize. Most often, ellipsoidal shapes are assumed where particles are taken to be horizontally aligned (Wu and Weinman, 1984) or to tumble slightly (Prigent et al., 2001. That the presence of non-spherical ice particles is not negligible is evident from several studies where considerable polarization occurs over strongly convective clouds and extended stratiform areas (Petty and Turk, 1996; Prigent et al., 2001).

Cloud ice: Small cloud ice particles may be treated in the same fashion as cloud droplets, i.e., neglecting size distribution and shape effects. Ice permittivity models are compiled in Pullianinen et al. (1998), for example. Small ice particles behave radiometrically like cloud droplets with less liquid water content. However, at higher frequencies (> 200 GHz) both particle shape and size distribution become very important (Evans et al., 1998).

Melting particles: The modeling of melting particle properties depends on the meltwater fraction as a function of fall distance below the zero degree isotherm as well as particle permittivity as a function of the mixing of individual contributions from air, ice, and water along the melting process.

From simulations of the melting layer it became evident that the largest uncertainties are related to the unknown particle spectra above the freezing level including the size-density relationship. This seems more crucial for snowflakes than for graupel since these tend to be larger and less dense thus creating bigger radiometric effects. Associated with that is the unknown distribution of meltwater inside the particle during melting. Most permittivity models treat the particle in a static way throughout melting which does not reflect realistic conditions as observed in the few available experimental studies (e.g. Mitra et al., 1990). Thus a multiple-stage model is required which can provide a more accurate description of particle composition (and maybe shape) associated with a flexible permittivity approach adjusted to the local conditions. Despite these uncertainties, Bauer et al. (2000) and Olson et al. (2001) show significant contributions of melting particles to upwelling radiances which, if not accounted for in database generation, may lead to systematic overestimations of surface rainfall rate in stratiform clouds.

4. RADIATIVE TRANSFER

Plane-parallel approximations to radiative transfer are strictly valid only in the limit of the absorption mean free path of photons being significantly shorter than the horizontal inhomogeneities of the medium. Recent advances in numerical cloud modeling have made available a great deal of fine scale cloud hydrometeor information together with cloud structure during its evolution. Such information can now be used to study the effects of arbitrary horizontal distribution.

In general, the plane-parallel assumption suffers from two types of problems that must be analyzed separately. The first is a physical problem, the second is geometric. The main physical problem is that the radiation is "trapped" in each sub-cloud considered as an infinite layer. At low frequencies the radiation emitted inside the cloud is not allowed to exit the sides of the cloud and, consequently, in terms of TB's the plane parallel results are slightly warmer when emission processes dominate. At high frequencies, when there is scattering in the higher layers, the average brightness temperature is lower for the same reasons.

The geometry problems arise because plane-parallel radiative transfer can not deal with inhomogeneities in horizontal direction. This leads to a geometric shift in the cloud radiometric footprint with respect to the real cloud position. This problem is partly overcome by adjusting the columns of the plane parallel model to the viewing direction along the slant path (Roberti et al., 1994; Bauer et al., 1998); however, more information on the accuracy of this correction is needed when very intense precipitation events are simulated. The nature and extend of the problem was demonstrated by several researchers using simple box-type cloud models (e.g. Haferman et al., 1996). For the purpose of uncertainty estimation of radiative transfer modeling vs. retrieval errors, an intercomparison of available models applied to various plane-parallel cloud profiles, surface conditions, and observation angles has been carried out (Smith et al., 2002). Figure 3 shows the set-up of cloud profiles used for testing various radiative transfer models in terms of volume extinction coefficient, k, single scattering albedo, ω_o, and asymmetry parameter, g.

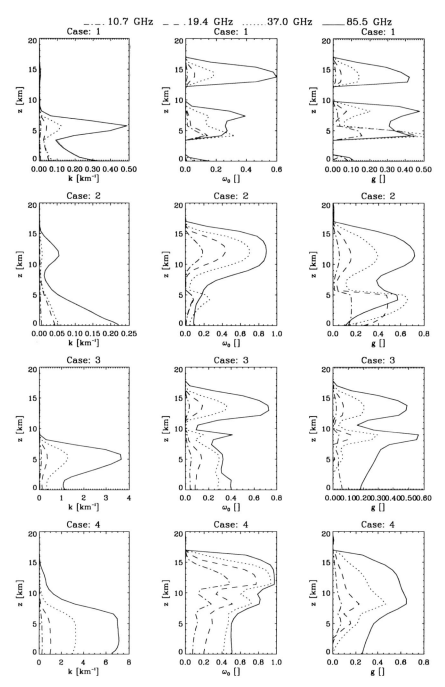

Figure 3: Cloud extinction (left column), scattering (middle column), and averaged phase function (right column) for 4 cloud profiles serving as input for radiative transfer model intercomparisons (10.7 GHz: dashed-dotted, 19.4 GHz: dashed, 37.0 GHz: dotted, 85.5 GHz: solid).

Precipitation modeling for inversion purposes 27

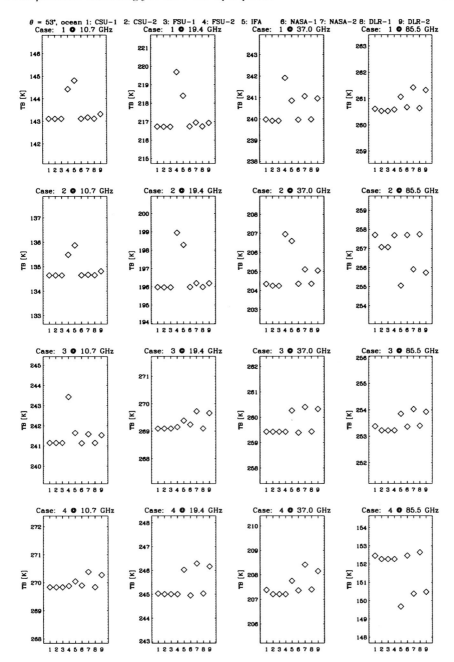

Figure 4: TB's over ocean at θ= 53.1°. CSU-1: CSU two-stream; CSU-2 as CSU-1 with δ-scaling; FSU-1: FSU two-stream, Fresnel surface; FSU-2 as FSU-1, Lambertian surface; IFA: discrete ordinate with Lambertian surface; NASA-1: two-stream, Fresnel surface; NASA-2: Monte-Carlo, Fresnel surface; DLR-1: matrix operator, Fresnel surface; DLR-2: two-stream, Fresnel surface.

Case 1 represents low water and ice contents, case 2 low water and large ice contents, case 3 low ice and large water contents, and case 4 large water and ice contents. All calculations were carried out at 10.7, 19.4, 37.0, and 85.5 GHz. Figure 4 shows TB's from all model simulations over a calm ocean surface at a zenith angle (θ) of 53.1°. Two features appear as a result of the treatment of surface reflection and scattering by hydrometeors. Those models (no. 4 and 5) which use the assumption of Lambertian surface reflection disagree with the others by several degrees (up to 20 K at 19.4 and 37.0 GHz for nadir incidence) which is reduced in the presence of optically thick clouds (case 4 and at 85.5 GHz). Secondly, those models which are multi-stream solutions to radiative transfer or Monte-Carlo models (no. 5, 7, 9) principally provide a more exact treatment of multiple scattering which becomes evident at large ice contents (cases 2 and 4) at 85.5 GHz. The δ-scaling (Joseph and Wiscombe, 1976) seems to be more efficient at lower zenith angles (not shown here) and can partially compensate for the lack of accuracy of the two-stream approximation models. From these intercomparisons, it may be suggested to either use multi-stream models with an accurate treatment of surface reflection or to use two-stream approximations including δ-scaling. In any case, effects of spatial cloud inhomogeneity have to be included.

5. DATABASE REPRESENTATIVENESS

Keeping all of the above radiative transfer aspects in mind, a test of database representativeness was carried out using satellite observations from the Tropical Rainfall Measuring Mission (TRMM) Microwave Imager (TMI) and Precipitation Radar (PR) (Kummerow et al. 1998). For this purpose, several cloud model simulations of tropical convection (see Table 1) were used in combination with a two-stream Eddington approximation model which accounts for three-dimensional effects to the first order (Bauer et al., 1998). Six different hydrometeor types were allowed with fixed (i.e. cloud model inherent exponential) particle size distributions including a melting layer (Bauer, 2001a). Satellite observation geometry and imaging specifications were accounted for by a frequency dependent spatial resolution together with variable satellite overpass directions and a nominal zenith angle of 52.8°. All individual simulations were combined using TRMM data to provide an optimized 'merged' database (Bauer, 2001b).

To compare with observations, TB's from 290 orbits of TMI data were analyzed. Figure 4 summarizes the statistics of all simulations and the observations. Figure 4a shows maximum, minimum, and mean TB's for all TMI channels. UW-MS/HC-3, GCE/TC-3, and Meso-NH/TC-1 represent more intense systems than GCE/TC-1 and JCMM/TC-1, i.e., the dynamic range of TB's is larger, that is 10.65 GHz TB's are higher due to more volume emission while 85.5 GHz TB's are lower as a result of more volume scattering. The observations contain as much variability as the more intense simulations. 85.5 GHz TB's as low as 100 K are simulated.

Table 1: Cloud model simulations in use.

Model	Reference	Cloud System	Resolution
Goddard Cumulus Ensemble Model, GCE-GT-2	Tao and Simpson (1989)	squall line, GATE	1.5 km
Goddard Cumulus Ensemble Model, GCE-TC-1	Tao and Simpson (1993)	squall line, TOGA-COARE	1.0 km
Goddard Cumulus Ensemble Model, GCE-TC-3	Tao and Simpson (1993)	squall line, TOGA-COARE	3.0 km
Univ. Wisconsin Modeling System, RAMS-HC-3	Tripoli (1992)	hurricane 'Gilbert'	3.3 km
Large Eddy Model UK Met. Office, JCMM-TC-1	Swann (1998)	squall line, TOGA-COARE	1.25 km
Large Eddy Model Meteo-France, Meso-NH-TC-1	Lafore et al. (1998)	squall line, TOGA-COARE	1.25 km

Comparing the merged database with the observations shows that the minimum 21.3 GHz TB's are higher as a consequence of the high water vapor burden in the simulations. The range of 10.65 (v, h) and 19.35 GHz (v) TB's is very similar while the largest differences occur comparing 37.0 (v,h) and 19.35 GHz (h) minima. This indicates that the full scale of observable conditions are not fully represented. Important is however, that the mean values and the range between means and maxima compare well because those represent the cloudy portion of the signal.

More insight is provided when empirical orthogonal functions (EOF's) are analyzed. Figure 4b presents the eigenvectors of the first three EOF's when all nine TMI channels are included while Figure 4c shows the same for the first two EOF's when only the lower seven channels are used. For each EOF the fractional explained variance is given in percent. In the first case, roughly 97-98% of total variability is explained by the first three EOF's.

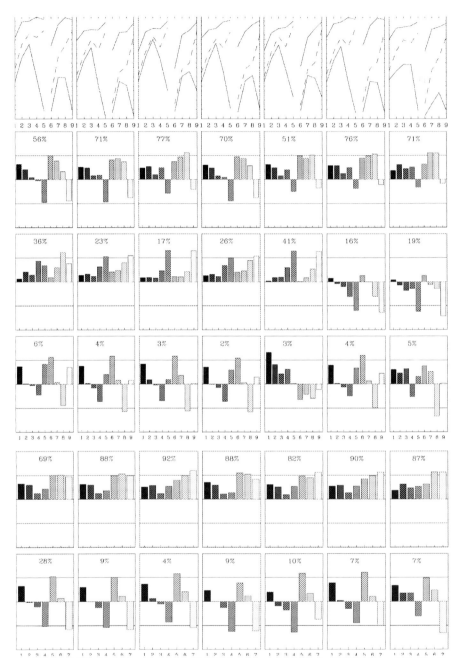

Figure 5: Minimum, maximum (solid line), and average (dashed) TB's from individual, merged simulations and TMI-PR observations (a). Channel numbers 1-9 refer to 10.65v, 19.35v, 21.3v, 37.0v, 85.5v, 10.65h, 19.35h, 37.0h, 85.5h. Figs. (b) give eigenvectors of first three EOFs as well as percentiles of explained variance. Figs. (c) show corresponding eigenvectors for seven channel EOFs with channel numbers 1-7 referring to 10.65v, 19.35v, 21.3v, 37.0v, 10.65h, 19.35h, 37.0h.

The same holds for the first two EOF's when the 85.5 GHz channels are excluded. For less intense cloud systems, the first EOF represents cloud emission for the 9-channel database. With increasing precipitating water and ice contents, the first two EOF's depend with similar sensitivity on integrated water and ice paths. The stronger systems in UW-NMS/HC-3, GCE/TC-3, and MesoNH/TC-1 show large positive values at low frequencies and large negative values at high frequencies in the eigenvector of the first EOF.

Thus, the first EOF expresses a strong frequency and a moderate polarization dependence. It therefore represents the major response to rain liquid water with positive response (emission) at lower frequencies, balanced response (emission and scattering) at 37.0 GHz, and negative response (scattering) at 85.5 GHz. Generally, the observations show less scattering and emission thus weaker systems. At 37.0 GHz there is strong positive response but less negative sensitivity at 85.5 GHz.

The second EOF's of the 9-channel databases always contain positive eigenvectors with increasing weight per frequency but little to no polarization effect. This may be interpreted as a sensitivity to cloud ice above opaque raincloud layers. Lower frequencies (10.65, 19.35, 21.3 GHz) contribute little. Very obvious is the difference to observations where the signs are negative but the relative magnitudes are the same. This results from remaining scattering not captured by the first EOF where scattering was less represented than in the simulations. Important to notice is that after database merging, similar eigenvectors are obtained from both simulations and observations.

In the 7-channel database, scattering features are generally reduced thus all eigenvector elements in the first EOF are positive. The EOF is sensitive to polarization and all window channels (10.65, 19.35, 37.0 GHz) have almost similar weight. Magnitude and polarization-difference of the eigenvector elements at 37.0 GHz indicate the representation of thinner clouds in the simulation. For example, GCE/TC-1 and JCMM/TC-1 show larger differences in 37.0 GHz polarization-difference than the other simulations. The second EOF shows a similar picture than the first EOF of the 9-channel databases. Thus the interpretation is similar, i.e., a strong response to total rain liquid water path in more convective areas. Here, the influence of the merging process is less obvious; however, simulations and observations show very similar patterns.

A better performance by individual models may indicate a better cloud model parameterization of hydrometeor microphysics; however, this intercomparison is too general to allow general conclusions on model quality. Another issue is the estimation of the non-uniqueness of the TB-cloud profile relationship which is one of the major retrieval error sources.

This was also evaluated using colocated TMI and PR measurements and comparing the rain liquid water content averaged over the larger TMI effective fields of view as well as their standard deviations between simulations and observations (Bauer et al., 2001; not shown here). Apart from the fact that simulations and observations showed very similar distributions, standard deviations in surface rain rate between 30-100% per brightness temperature vector were obtained. These magnitudes have to be considered when total errors of rainfall retrievals are estimated. This also leads to the requirement that information other than the TB-vector at the actual location may help to narrow down the uncertainty, e.g., by criteria using spatial TB-variability for stratiform-convective separation (e.g. Hong et al. 1999).

6. CONCLUSIONS

The generation of simulated observations for precipitation retrieval algorithm training is extremely difficult because optical properties of surfaces, atmosphere, and hydrometers have to be considered. Of these, the complex precipitation microphysics requires a synergy of models for particle spectra, composition and shape, cloud geometry, accurate radiative transfer as well as a simulation tool with the imaging specifications of the targeted radiometer. The errors associated with these models are not well known so that integral modeling errors are unknown. These can only be estimated indirectly, for example by statistically comparing simulated and observed radiances.

There are many studies dealing with individual problems in this context. In these studies, however, all other problems depicted in Figure 2 are treated in an oversimplified manner. Consequently, a real synergy of state-of-the-art models including *all* problems is still missing. Thus current retrieval databases always suffer from at least a few major oversimplifications.

As a guideline, a ranking of problem importance to be addressed may be:
1. three-dimensional distribution of hydrometeor concentrations,
2. radiative transfer model and radiometer imaging simulation,
3. surface emissivity model and surface state variability,
4. particle size spectra,
5. particle composition,
6. particle shape and atmospheric absorption.

From this, it is evident that there is a high demand for more and better cloud model simulations serving as input for retrieval databases. Therefore a strong interaction between the precipitation retrieval and cloud modeling communities is needed.

7. REFERENCES

Bauer, P. and P. Schlüssel, 1993: Rainfall, total water, ice water and water-vapour over sea from polarized microwave simulations and SSM/I data, J. Geophys. Res., 98, 20737-20759.

Bauer, P., L. Schanz, and L. Roberti, 1998: Correction of three-dimensional effects for passive microwave retrievals of convective clouds. J. Appl. Meteor., 37, 1619-1632.

Bauer, P., A. Khain, A. Pokrovsky, R. Meneghini, C. Kummerow, and J.P.V. Poiares Baptista, 2000: Combined cloud-microwave radiative transfer modeling of stratiform rainfall. J. Atmos. Sci., 57, 1082-1104.

Bauer, P., 2001a: Microwave radiative transfer simulation in clouds: Including a melting layer in cloud model bulk hydrometeor distributions. Atmos. Res., 57, 9-30.

Bauer, P., 2001b: Over-ocean rainfall retrieval from multi-sensor data of the Tropical Rainfall Measuring Mission (TRMM)-Part I: Development of inversion databases. J. Atmos. Ocean. Tech., 18, 1315-1330.

Bauer, P., P. Amayenc, C.D. Kummerow, and E.A. Smith, 2001: Over-ocean rainfall retrieval from multi-sensor data of the Tropical Rainfall Measuring Mission (TRMM)-Part II: Alghorithm inplementation. J. Atmos. Ocean. Tech., 18, 1838-1855.

Czekala, H. and C. Simmer, 1998: Microwave radiative transfer with non-spherical precipitating hydrometeors. J. Quant. Spec. Rad. Trans., 60, 365-374.

Chanzy, A. and J.-P. Wigneron, 1999: Microwave emission from soil and vegetation. In: Radiative transfer models for microwave radiometry (Ed: C. Maetzler), COST-712 final report of project 1, University of Bern, pp. 174.

Deblonde, G., 1999: Variational assimilation of SSM/I total precipitable water retrievals in the CMC analysis system. Mon. Wea. Rev, 127, 1458-1476.

Evans, K.F., S.J. Walter, A.J. Heymsfield, and M.N. Deeter, 1998: Modeling of submillimeter passive remote sensing of cirrus clouds. J. Appl. Meteor., 37, 184-205.

Eymard, L., P. Sobieski, D. Lemaire, E. Obligis, S. English, and T. Hewison, 1999: Ocean surface emissivity modeling. In: Radiative transfer models for microwave radiometry (Ed: C. Maetzler), COST-712 final report of project 1, University of Bern, pp. 174.

Eyre, J.R., G.A. Kelly, A.P. McNally, E. Andersson, and A. Persson, 1993: Assimilation of TOVS radiance information through one-dimensional variational analysis. Q. J. R. Meteorol. Soc., 119, 1427-1463.

Ferrier, B.S., 1994: A double-moment multiple-phase four-class bulk ice scheme. Part I: Description. J. Atmos. Sci., 51, 249-280.

Haddad, Z., E. Smith, C. Kummerow, T. Iguchi, M. Farrar, S. Durden, M. Alves, and W. Olson, 1997: The TRMM 'day-1' radar/radiometer combined rain-profiling algorithm. J. Meteor. Soc. Japan, 75, 799-808.

Haferman, J., E. Anagnostou, D. Tsintikidis, W. Krajewski, and T. Smith, T., 1996: Physically based satellite retrieval of precipitation using a 3-d passive microwave radiative transfer model. J. Atmos. Ocean. Tech., 13, 832-850.

Hong, Y., C. Kummerow, and W. Olson, 1999: Separation of convective and stratiform precipitation using microwave brightness temperatures. J. Appl. Meteor., 38, 1195-1213.

Joseph, J.H. and W.J. Wiscombe, 1976: The Delta-Eddington approximation for radiative flux transfer. J. Atmos. Sci., 33, 2452-2459.

Kummerow, C., W. Olson, and L. Giglio, 1996: A simplified scheme for obtaining precipitation and vertical hydrometeor profiles from passive microwave sensors. IEEE Trans. Geosci. Remote Sens., 34, 1213-1232.

Kummerow, C., W. Barnes, T. Kozu, J. Shiue, and J. Simpson, 1998: The Tropical Rainfall Measuring Mission (TRMM) sensor package. J. Atmos. Ocean. Tech., 15, 809-817.

Lafore, J.-P. and co-authors, 1998: The Meso-NH atmospheric simulation system. Part I: Adiabatic formulation and control simulations. Ann. Geophys., 16, 90-109.

Liebe, H., P. Rosenkranz, and G. Hufford, 1992: Atmospheric 60 GHz oxygen spectrum: New laboratory measurements and line parameters. J. Quant. Spec. Rad. Trans., 48, 629-643.

Lipton, A.E., M.K. Griffin, and A.G. Ling, 1999: Microwave transfer model differences in remote sensing of cloud liquid water at low temperatures. IEEE Trans. Geosci. Remote Sens, 37, 620-623.

Liu, G. and. J.A. Curry, 1992: Retrieval of precipitation from satellite microwave measurement using both emission and scattering. J. Geophys. Res., 97, 9959-9974.

McKague, D., K.F. Evans, and S. Avery, 1998: Assessment of the effects of drop size distribution variations retrieved from UHF radar on passive microwave remote sensing of precipitation. J. Appl. Meteor., 37, 155-165.

Mitra, S.K., O. Vohl, M. Ahr, and H.R. Pruppacher, 1990: A wind tunnel and theoretical study of the melting behavior of atmospheric ice particles. IV: Experiment and theory for snow flakes. J. Atmos. Sci., 47, 584-591.

Mugnai A. and E.A. Smith, 1988: Radiative transfer to space through a precipitating cloud at multiple microwave frequencies. Part I: Model description. J. Appl. Meteor., 27 1055-1073.

Olson, W., P. Bauer, N. Viltard, D. Johnson, and W.-K. Tao, 2001: A melting layer model for passive / active microwave remote sensing applications-Part I: Model formulation and comparison with observations. J. Appl. Meteor., 40, 1145-1163.

Panegrossi, G. and co-authors, 1998: Use of cloud model microphysics for passive microwave-based precipitation retrieval: Significance of consistency between model and measurement manifolds. J. Atmos. Sci., 55, 1644-1673.

Petty, G.W. and J. Turk, 1996: Observed multichannel microwave signatures of spatially extensive precipitation in tropical cyclones. Proceedings 8th Conf. Sat. Meteorol. Ocean., Atlanta GA, 291-294.

Petty, G.W., A. Mugnai, and E.A. Smith, 1994: Reverse Monte-Carlo simulations of microwave radiative transfer in realistic 3-d rain clouds. Proc. 7th AMS Conf. Sat. Meteor. Ocean., 185-188.

Prigent, C., J.R. Pardo, M.I. Mishchenko, and W.B. Rossow, 2001: Microwave polarized scattering signatures in clouds: SSM/I observations with radiative transfer simulations. J. Geophys. Res., 106, 28243-28258.

Pullianinen, J., K. Tigerstedt, W. Huining, M. Hallikainen, C. Maetzler, A. Wiesmann, and C. Wegmueller, 1998: Retrieval of geophysical parameters with integrated modeling of land surfaces and atmospheres (models / inversion algorithms. Final Report ESA/ESTEC, Noordwijk, The Netherlands, Contract No. 11706/95/NL/NB(SC), pp. 274.

Roberti., L., J. Haferman, and C. Kummerow, 1994: Microwave radiative transfer through horizontally inhomogeneous precipitating cloud. J. Geophys. Res., 99, 16707-16718.

Rodgers, C.D., 1976: retrieval of atmospheric temperature and composition from remote measurements of thermal radiation. Rev. Geophys. Space Phys., 14, 609-624.

Schols, J.L., Weinman, J.A., Alexander, G.D., Stewart, R.E., Angus, L.J., and A.C.L. Lee, 1999: Microwave properties of frozen precipitation around a North Atlantic cyclone. J. Appl. Meteor., 38, 29-43.

Smith, E.A. and co-authors, 1998: Results of WetNet PIP-2 project. J. Atmos. Sci., 55, 1483-1536.

Smith, E.A., P. Bauer, F.S. Marzano, C.D. Kummerow, D. McKague, A. Mugnai, and G. Panegrossi, 2002: Intercomparison of microwave radiative transfer models for precipitating clouds. IEEE Trans. Geosci. Remote Sens., in press.

Profile Retrieval Estimation Techniques

ED R. WESTWATER
Cooperative Institute for Research in Environmental Sciences (CIRES), University of Colorado/NOAA Environmental Technology Laboratory,Boulder, Colorado

Key words: inverse problems, radiometric remote sensing, optimal estimation, regularization methods, iterative techniques

Abstract: Retrieval techniques, applicable to problems where measurements have to be blended, satisfy an equation that can be approximated by a Fredholm integral equation of the first kind. A variety of inversion techniques, that have been used in radiometric remote sensing of temperature and moisture profiles, are outlined to solve this equation focusing on the wide variety of supplementary information that may be used in addition to radiance information. Techniques, such as regularization and iterative methods, are introduced because of their simplicity and explicit use of an initial guess. Methods based on Backus-Gilbert and Kalman approaches are shown for diagnosis problems. Statistical techniques are illustrated emphasizing the use a priori information to constrain the solution. Final remarks give the current scenario on this subject.

1. INTRODUCTION

Currently remote-sounders operating from both ground- and satellite-based platforms provide excellent data for temperature, water vapor, and cloud liquid profiles. However, neither for temperature nor for moisture is any system completely adequate for routine operational or research purposes. On the other hand, there are excellent possibilities that a proper blend of existing instruments and data could satisfy a variety of requirements. In this summary, I sketch some of the commonly used techniques that can be used to blend data from diverse sources. Although many of the techniques have been applied to radiometric retrieval problems, some, at least, have a more general application. The techniques are applicable to problems when at least some of the measurements to be blended satisfy an equation that can be approximated by a Fredholm integral equation of the first kind (Twomey 1977). This equation in its discrete form is written

$$g_\varepsilon = Af \tag{1}$$

where g_ε is a vector composed of n measurements, $\mathbf{f}$ is an m-vector whose components are the profile that we want to determine, A is an $n \times m$ matrix relating the measurements to the unknown, and the n-vector ε explicitly denotes that the measurements have an unknown error component that will affect the solution to some degree. For mildly nonlinear problems, a perturbation form of (1) is frequently used as the basis of subsequent iterations. Retrieval algorithms that require calculations of A for their implementation are called "physical." In addition to the books by Twomey (1977), Tikonhov and Arsenin (1977),Deepak (1977),and Deepak et al. (1989); an excellent review article discussing techniques of solving (1) was written by Rodgers (1976).

Data to be blended can have completely different characters (Westwater, 1997). For example, we may want to blend ground- and satellite-based radiance data that contain information on both temperature and moisture, Global Positioning System (GPS) soundings of total water content, Radio Acoustic Sounding System (RASS) soundings of virtual temperature profiles that extend only partly through the region of interest, and perhaps a sounding from a radiosonde that was launched several hours previously. Each of these sources of information can be expressed by (1), where, for example, A in radiometric remote sensing describes the so-called weighting functions (Westwater 1993), A in Global Positioning System sounding gives approximately equal weight to water vapor density at each altitude, and A for Radio Acoustic Sounding Systems would be a partially filled diagonal matrix. Another source of useful information from ground-based active sensors, such as a cloud lidar or radar, is a measurement of cloud height and/or thickness. As a general rule, whatever data that we blend should be consistent with each other to within the known accuracy of the measurements. In addition, since the inverse problem is generally ill posed, a priori knowledge and other supplementary knowledge is essential to deriving useful solutions of (1).

2. REGULARIZATION

This technique was developed in the early 1960s and is discussed thoroughly in the books by Tikhonov and Arsenin (1977) and Twomey (1977). This technique introduces a controlled smoothing between the solution of (1) and an a priori guess. The smoothing is controlled by a regularization parameter γ which balances a priori and measurement information (Westwater and Strand 1968).

This technique is simple and widely used in satellite thermal sounding (Rao et al. 1990) and is sometimes called the "minimum information" solution (Fleming and Smith 1971). With this technique, for example, first guess information from a meteorological forecast, or a climatological average, can readily be used in the solution Fleming (1977). This technique, as well as many generalizations, minimizes a quadratic form Q(**f**) that contains two strictly positive terms

$$Q(f) = [(Af - g_r)'(Af - g_r)] + \gamma[(f - f_0)'(f - f_0)] \quad (2)$$

where f_0 is the initial guess. Minimizing (2) yields

$$f = f_0 + A'(AA' + \gamma I)^{-1}(g_r - Af) \quad (3)$$

A thorough discussion of the choice of (is given by Twomey (1977) and the relationship between (and signal-to-noise ratio is discussed by Westwater and Strand (1968). An example of temperature profile retrieval accuracy using this method is shown in Fig. 1.

3. ITERATIVE TECHNIQUES

Iterative techniques, in which a solution **f** is approached by successive approximations, were pioneered by Chahine (1970) and Landweber (1951) and have been widely applied (Smith 1983; Chedin et al. 1985). Strand (1974) showed that there was a close connection between iterative techniques and regularization methods, with a relationship between the smoothing parameter γ and the number of iterations. The general form of an iterative technique is

$$f_k = f_{k-1} + DA'(g_r - Af_{k-1}), \quad k = 1, 2, 3..... \quad (4)$$

where D is a specified matrix. In particular, Strand (1974) showed that the choice of D

$$D = (A'A + \gamma I)^{-1} \quad (5)$$

with starting vector f_0 would give (3) for the first iteration.

4. BACKUS-GILBERT RESOLUTION ANALYSIS

A method of analyzing the spatial resolution of passive remote sounding systems was published by Backus and Gilbert (1970). In this method, a linear combination of weighting functions is used to form a so called avera-

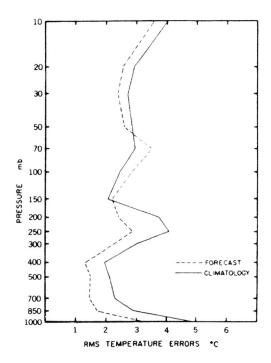

Figure 1. Rms temperature retrieval errors between 139 radiosonde profiles and regularization retrievals when forecasts and climatology are used as initial profiles in satellite retrieval simulations. After Fleming (1977).

ging kernel $A(y_0, y)$, where y_0 is a fixed spatial coordinate and y is a running spatial variable. The spread function at y_0 can be used to estimate the spatial resolution and trade off between spread and measurement error is a useful system characterization. To describe this method, we change from the discrete matrix equation (1) to its continuous integral equation form. The basic integral equation with perfect measurements g_i, unknown function $f(y)$, weighting function $W_i(y)$, and measurement error ε_i is written

$$(g_i)_\varepsilon = g_i + \varepsilon_i = \int W_i(y) f(y) dy + \varepsilon_i \qquad i = 1, 2, \ldots, n \tag{6}$$

where W_i is the weighting function or kernel to the Fredholm integral equation. The basic idea of Backus and Gilbert (1970) (BG) is to construct averaging kernels which can be used to construct averages of $f(y)$ in the neighborhood of some point y_0. Let $A(y_0, y)$ be constructed from some linear combination of $W_i(y)$, with coefficients $a_i(y_0)$

$$A(y_0, y) = \sum_i^n a_i(y_0) W_i(y) \tag{7}$$

Here, BG only consider unimodular averaging kernels:

Profile retrieval estimation techniques

$$\int A(y_0, y) dy = 0 \tag{8}$$

Thus linear combinations of the measurements represent weighted averages of f that, with suitable choices of $a_i(y_0)$, may be centered about y_0

$$\langle\langle f(y) \rangle\rangle_{y_0} = \int A(y_0, y) f(y) dy + \sum_{1}^{n} a_i(y_0) \varepsilon_i \tag{9}$$

We have written $\langle\langle \cdot \rangle\rangle$ to indicate spatial, in contrast to statistical ensemble, average. The Backus-Gilbert technique of analyzing resolution at y_0 defines the spread $s(y_0)$

$$s(y_0) = A'(y_0) S(y_0) A(y_0) \tag{10}$$

as determined from the spread tensor $S_{ij}(y_0)$

$$S_{ij}(y_0) = 12 \int W_i(y_0) W_j(y_0)(y - y_0)^2 dy \tag{11}$$

subject to the constraint that

$$a(y_0)' u = 1 \tag{12}$$

where

$$u_i = \int W_i(y) dy \tag{13}$$

We also must consider the error variance at y0, given the measurement error covariance matrix E, as

$$Var\left(\langle\langle f(y) \rangle\rangle_{y_0}\right) \equiv e^2(y_0) = A(y_0)' E(y_0) \tag{14}$$

Backus-Gilbert discuss the tradeoff between spread and error variance and conveniently parameterize the tradeoff by the quadratic form $Q(\theta)$, $0 \leq \theta \leq (\pi/2)$, where

$$Q(\Theta) = s(y_0) + w\varepsilon^2(y_0) \sin(\Theta) \tag{15}$$

The scaling parameter w is used to insure that as θ varies over its range, that the variations in the tradeoff curve are distributed uniformly with θ Without care in w, most of the variation in the tradeoff curve will occur near the extremes θ= 0 and θ= π/2. We minimize (14), subject to the constraint (7) to obtain the equations for coefficients $a_i(y_0)$

$$a(y_0) = \frac{1}{(u' R_{y_0} u)} R_{y_0}(\Theta)^{-1} u \tag{16}$$

where

$$R_{y_0}(\Theta) = S(y_0)\cos(\Theta) + wE(y_0)\sin(\Theta) \tag{17}$$

Both the minimum and maximum $s(y_0)$ and $e^2(y_0)$ occur at the ends of the tradeoff curve:
$s_{min} = s(0)$, $s_{max} = s(\pi/2)$, $e^2_{min} = e^2(\pi/2)$, and $e^2_{max} = e^2(0)$

5. BACKUS-GILBERT METHOD USING A PRIORI STATISTICS

As discussed by Rodgers (1976), information other than radiation measurements can be incorporated into the spread-error analysis. For example, we frequently have a priori climatological statistics, a forecast from a numerical model, or some equivalent source of information. For simplicity, we will assume that we have a priori statistics determined from a past collection of radiosondes. To use such information, the error characteristics of the source of information is needed. Here, we assume that the error characteristics of **f** are completely described by its covariance matrix over some representative ensemble of profiles. Let <**f**> be the ensemble average of **f** and $E(\cdot)$ be the expected value operator over this ensemble. The covariance matrix of **f**, S_f is given by

$$S_f = E(f - \langle f \rangle)(f - \langle f \rangle)^T \tag{18}$$

Conceptually, we may regard <**f**> as a measurement of **f**, with a spatial resolution given by the characteristics of the radiosonde sensor, and an error covariance given by S_f. Thus, we write

$$\langle\langle\langle f(y)\rangle\rangle\rangle_{y_0} = \int W(y_0, y)\langle f(y)\rangle dy \tag{19}$$

where we assume that the radiosonde weighting function $W(y_0,y)$ is a rectangular function centered in some set of reporting levels. The methods of sections 4 and 5 have been used by Westwater et al. (2000) to evaluate the accuracy and vertical resolution of multi-angle and multi-frequency ground-based radiometric systems.

6. LINEAR STATISTICAL RETRIEVAL

This method is used to incorporate a priori statistics, including means and covariance matrices of f and ε into the retrieval process (Westwater and Strand 1968; Rodgers 1970; Rodgers 1976). In complete analogy to (2), and assuming Gaussian statistics for both the profile vector f and the measurement error e, a maximum likelihood estimate for the profile is obtained by minimizing the quadratic form

$$Q(f) = (Af - g_e)^T S_\varepsilon^{-1}(Af - g_e) + (f - \langle f \rangle)^T S_f^{-1}(f - \langle f \rangle) \tag{20}$$

with respect to **f**. Minimizing (20) yields the linear statistical retrieval equation (LSRE)

$$\hat{f} \langle f \rangle + S_f A^T \left(A S_f A^T + S_\varepsilon \right)^{-1} \left(g_e - A \langle f \rangle \right) \tag{21}$$

where is the estimate of the profile vector **f** and here < > designates ensemble averaging over a representative collection of profiles and measurements. As seen above, the equations for regularization and iteration can be cast in a form similar to (21). For a linear problem with Gaussian statistics, the LSRE is optimum and frequently, retrieval problems are nearly linear and obey Gaussian statistics. However, a more serious limitation is one of obtaining representative statistics because, in practice, the means and covariance matrices of f are estimated from a history of radiosondes. Because operational radiosondes are usually taken twice-a-day, acquisition of representative statistics at non synoptic times is problematic. A slightly more general form of (21) is written as

$$\hat{f}' = E\left(f' g_e'^T\right) E\left(g_e' g_e'^T\right)^{-1} g_e' \tag{22}$$

where the primed quantities refer to departures from the ensemble mean. A very useful feature of statistical methods is that they immediately yield an estimate of profile retrieval error; i. e., the covariance matrix of the difference between the profile and its estimate

$$S_{f-\hat{f}} = S_f - S_{f,g_e} S_{g_e}^{-1} S_{g_e,f} \tag{23}$$

where

$$S_{f,g_e} \equiv E\left((f - \langle f \rangle)(g_e - \langle g_e \rangle)^T \right) \tag{24}$$

The statistical methods of this section all require a forward model, in which the measured quantity g_e can be calculated from **f**. Rodgers (1990) discussed how the error covariance matrix, Se is composed of both instrument and forward modeling errors. An example of the application of (23) to retrieval error prediction is shown in Fig. 2.

7. REGRESSION METHODS

In this method, if an a posteriori ensemble of simultaneous and collocated remote sensor and in situ measurements is available, then multivariate linear

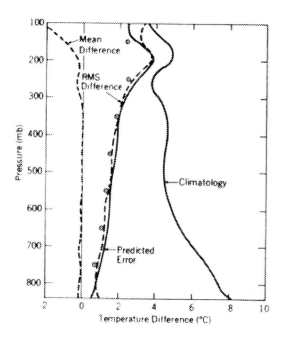

Figure 2. Predicted temperature profile retrieval errors and climatological variations (solid curve), observed differences between retrievals and radiosondes (dashed curves), and rms differences of layer-averaged temperatures for 100-mbar cire layers (circles). Results based on 6-channel ground-based radiometer data taken at Denver, Colorado, USA. After Westwater (1993).

regression (i., e., equation (22)) is used to provide a set of matrix coefficients that can be used in subsequent applications. This technique was used operationally by the United States National Environmental Satellite Distribution and Information Service for several years (Phillips et al. 1979). It has the advantage of incorporating instrumental biases into retrieval coefficients and does not require solution of the forward problem, i. e., that of calculating the measurements, given the profile. As such, it is called a nonphysical method, since no understanding of the underlying physics is required to construct the profile retrieval coefficients.

8. INVERSE COVARIANCE WEIGHTING

This method is particularly appealing when combining estimates of a profile f from two or more independent sources of information whose error characteristics are known (Rodgers 1976). For example, if we had a thermal retrieval from a satellite observing system f_1, whose accuracy was described by a covariance matrix S_1 and a retrieval from an operational forecast f_2,

Profile retrieval estimation techniques 43

with covariance matrix S_2, then the blended profile f_C' could be determined by

$$\hat{f}' = \left(S_1^{-1} + S_2^{-1}\right)^{-1}\left(S_1^{-1}f'_1 + S_{21}^{-1}f'_2\right) \qquad (25)$$

The method is readily generalized to three or more independent systems. Equation (25) was used by Westwater et al. (1984) and by Schroeder et al. (1991) to combine independent ground- and satellite-based retrievals of temperature profiles.

9. NEURAL NETWORK INVERSION

The application of neural networks to retrieval of temperature profiles from ground-based radiometric measurements was considered by Churnside et al. (1994), and neural networks have been applied to radiometric retrievals (Moteller et al. (1995); Del Frate, F. and G. Schiavon, (1998). The application of this method requires a training set, say, in the form of radiances and profiles, and an algorithm to determine the weights of a hidden layer of neurons relating the input (radiances) to the output (profile) vectors. Neither linearity nor a definite physical relationship is required. Churnside et al. (1994) found that over an ensemble of profiles, the rms error in profile retrieval was basically the same as that arising from linear statistical inversion. However, unusual and difficult cases seemed to be retrieved somewhat better. Neural networks might also be an ideal way to combine completely disparate sources of information, such as radiance and cloud-base height, or radiance and radar reflectivity. With the rapid development of neural network methods, their application to profile retrieval is a promising area of research.

10. KALMAN FILTERING

Applications of Kalman Filtering (KF) to derive atmospheric variables have been reported by Ledsham and Staelin (1978) and Wang et al. (1983) for space-based remote sensing and by Askne and Westwater (1986), Basili et al. (1981) and Han et al. (1997) for ground-based sensing. A general description of this technique is given by Gelb (1988) and by Brown and Hwang (1997). In discrete KF, a state vector s_i describes the state of a system at time step i. The state vector, in turn, evolves in time as a linear dynamic system according to a given evolution model. A measurement vector z_i contains measurements for the system. From z_j at all previous time increments (j = 0, 1 ···, i) KF estimates s_i. Here, s_i could be the f_i that satisfies the integral equation (1) or a more general parameter vector. The merit of this technique is that the estimation uses not only current measurements but also a history of measurements. In addition, covariance matrices describing the error in estimating s_i are available for each i. The most difficult aspect in the construction of KF equations is that of developing a reasonable model to describe the temporal evolution of the state vector. This difficulty is mitigated by the flexibility and power in choosing z_i ; e. g., z_i at one time could be a radiance, at another time a radiosonde profile, and at yet another time the output of a numerical forecast model.

11. HYBRID METHODS

When data are combined from different sources, it is sometimes advantageous to apply different retrieval methods to each of the datasets and then to combine the resulting retrievals. Inverse covariance weighting (section 8) is one way of combining. Another way method was suggested by Stankov (1996), who combined ground- and satellite-based profile retrievals by a hybrid method.

Ground-based temperature and humidity retrievals from RASS and microwave radiometers were combined with in situ measurements from aircraft by linear statistical inversion. The resulting retrieval was then used as a first guess for the physical iterative International TOVS [TIROS-N Operational Vertical Sounder] Processing Package (ITPP; see Rao et al. 1990). Chedin et al. (1985) used the combination of a forecast and an a priori ensemble of profiles to determine a first guess for a physical-iterative retrieval. A similar method was used by Han and Westwater (1995), who used a statistical retrieval as a first guess for an physical iterative method. As another example, Frisch et al. (1995) combined cloud radar and microwave radiometer data by using the integrated liquid derived from the radiometer, together with an assumed size distribution, to derive profiles of cloud liquid density from zero, first, and second moments of Doppler radar data. One could also include "classification" methods (McMillin 1991; Udstrom and Wark 1985, Thompson et al., 1985) in the hybrid category. In another application to moisture (vapor and liquid) retrievals, the retrieved profiles were constrained to be positive functions (Tatarskii et al.,1996). In general, almost any of the methods in sections 2-10 could be combined in a hybrid manner. Thus, it is apparent that a wide range of techniques are available to combine remote sensing data, but the choice of which one depends somewhat on the amount of a priori information, including an initial guess, that one has available (Westwater, 1997).

12. CONCLUDING REMARKS

In this chapter, we have outlined a variety of inversion techniques that have been used in radiometric remote sensing of temperature and moisture profiles. None of the techniques is clearly superior to the others, mainly because of the wide variety of supplementary information that may be used in addition to radiance information. Some techniques, such as regularization or iterative methods, are commonly used because of their simplicity and explicit use of an initial guess. Methods based on the ideas of Backus and Gilbert (1970) and extended by Rodgers (1976), are useful for diagnosis. Statistical techniques use a priori information to constrain the solution. Use of advanced techniques, such as neural networks, still require adequate training sets, but disparate information, such as radiance data, cloud height data, radar reflectivities, etc., can all be used as an input to the algorithm. A technique of great promise is Kalman Filtering, in which a dynamical model of the atmospheric state is coupled with measurements. Finally, substantial promise is being shown, especially in the use of satellite radiance data, by radiance assimilation methods (English, 2002). Because of the impact of combined model-sensor methods, previous ill-posed inverse problems have been replaced by properly-posed ones, and increased attention is now directed towards improvements in forward models of radiative transfer, especially in the presence of clouds.

13. REFERENCES

Askne, J. and E.R. Westwater, "A Review of Ground-Based Remote Sensing of Temperature and Moisture by Passive Microwave Radiometers," IEEE Trans. Geoscience and Remote Sensing, Vol. G3-24, 340-352, (1986).

Backus, G. E. and J. F. Gilbert, "Uniqueness in the inversion of inaccurate gross earth data", Phil. Trans. Roy. Soc. London, Ser. A 266, 123-192 (1970).

Basili, P. P. Ciotti and D. Solimini, "Inversion of ground-based radiometric data by Kalman filtering", Radio Sci., 16, 83-91 (1981).

Brown, R. G. and P. Y. C. Hwang, Introduction to random signals and applied Kalman Filtering, J. Wiley & Sons, New York, (1997)

Chahine, M. T., "Inverse problems in radiative transfer: determination of atmospheric parameters", J. Atmos. Sci., 27, 960-967 (1970).

Chedin, A., N. A. Scott, C. Wahiche, and P. Moulinier, "The Improved Initialization Method: A high resolution physical method for improved temperature retrievals from satellites of the TIROS-N series". J. Climate Appl. Meteor., 24, 128-143,(1985).

Churnside, J. H., T. A. Stermitz, and J. A. Schroeder, "Temperature profiling with neural network inversion of microwave radiometer data", J. Atmos. Oceanic Technol., 11, 105-109.(1994).

Deepak, A. (ed.), Inversion Methods in Atmospheric Remote Sounding. Academic Press, Inc., New York. (1977).

Deepak, A., H. E. Fleming, and J. S. Theon, (eds.), RSRM'87 Advances in Remote Sensing Retrieval Methods. A. Deepak Publishing, Hampton, VA. (1989).

Del Frate, F. and G. Schiavon, "A combined natural orthogonal functions/neural network technique for radiometric estimation of atmospheric profiles", Radio Science, 33, 405-410. (1998).

English, S., Atmospheric remote sensing by microwave and visible-infrared sensors Temperature profiles by microwave radiometry, Chapter 2 of this book.

Fleming, H. E. "Comparison of linear inversion methods by examination of the duality between iterative and inverse matrix methods", in Inversion Methods in Atmospheric Remote Sounding. A. Deepak (Ed.), Acad. Press (1977).

and W. S. Smith, "Inversion techniques for remote sensing of atmospheric temperature profiles". Fifth Symposium on Temperature, Washington, DC, Instrument Soc. of Amer., 2239-2250, (1971).

Han, Y., and E. R. Westwater, "Remote sensing of tropospheric water vapor and cloud liquid water by integrated ground-based sensors". J. Atmos. Oceanic Technol., 12, 1050-1059,(1995)

Han, Y., E. R. Westwater, and R. A. Ferrare, "Applications of Kalman filtering to derive water vapor from Raman lidar and microwave radiometers," J. Atmos. Oceanic Technol., 14, 480-487, (1997)

Landweber, L., "An iteration formula for Fredholm integral equations of the first kind," Amer. J. Math., 73, 615-624, (1951)

Ledskam, W. M.and D. H. Staelin, "An extended Kalman-Bucy filter for atmospheric temperature profile retrieval with a passive microwave sounder", J. Appl. Meteor. 17, 1023- 1033 (1978).

McMillan, L. M., "Evaluation of a classification method for retrieving
atmospheric temperatures from satellite measurements", J. Appl. Meteor., 30, 432-466 (1991).

Moteller, H. E., L. L. Strow, and L. McMillin, and J. A. Gualtieri, "Comparison of neural networks and regression-based methods for temperature retrievals". Appl. Opt., 34, 5390-5397., (1995)

Phillips, N. A., L. M. McMillin, D. Wark, and A. Gruber, "An evaluation of early operational temperature soundings from TIROS-N", Bull. Amer. Meteor. Soc., 60, 1188-1197.(1979).

Rodgers, C. D., "Retrieval of atmospheric temperature and composition from remote measurements of thermal radiation", Rev. Geophys. Space Phys., 14, 609-624 (1976).

Rodgers, C. D., "The vertical resolution of remotely sounded temperature profiles with a priori statistics", J. Atmos. Sci., 33, 707-709 (1976).

Rodgers, C. D., "Characterization and error analysis of profiles retrieved from remote sounding measurements,", J. Geophys. Res., 95, D5, 5587-5595. (1990)

Schroeder, J. A., E. R. Westwater, P. T. May, and L. M. McMillin, "Prospects for temperature sounding with satellite and ground-based RASS measurements", J. Atmos. Oceanic Technol., 8, 506-513. (1991).

Smith, W. L. ,"The retrieval of atmospheric profiles from VAS geostationary radiance observations", J. Atmos. Sci., 40, 2025-2035 (1983).

Strand, O. N. ,"Theory and methods related to the singu-lar-function expansion and Landweber's iteration for integral equations of the first kind", SIAM J. Numer. Anal. 11, 798-825 (1974).

Tatarskii, V. V., M. S. Tatarskaia and E. R. Westwater, "Statistical retrieval of humidity profiles from surface measurements of humidity, temperature and measurements of integrated moisture content", J. Atmos. and Oceanic Technol., 13, 165-174. (1996).

Thompson,O., M. Goldberg and D. Dazlich, "Pattern recognition in the satellite temperature retrieval problem", J. Climate Appl. Meteor. 24, 30-48 (1985).

Tikhonov, A. N., and V. Y. Arsenin, Solutions of Ill-Posed Problems. V.H. Winston and Sons, Wash., D.C. (1977).

Twomey, S., Introduction to the Mathematics of Inversion in Remote Sensing and Indirect Mea-sure-ments. Elsevier, New York. (1977).

Uddstrom, M. J., and D. Q. Wark, "A classification scheme for satellite temperature retrievals", J. Climate & Appl. Meteor., 24, 16-29 (1985).

Wang J. R., J. L. King, T. T. Wilheit, G. Szejwach, L. H. Gesell, R. A. Nieman, D. S. Niver, B. M. Krupp and J. A. Gagliano, "Profiling atmospheric water vapor by microwave radiometr" J. Climate Appl. Meteor., 22, 779-788, (1983).

Westwater, E. R. and O. N. Strand, "Statistical information content of radiation measurements used in indirect sensing", J. Atmos. Sciences, 25, 750-758 (1968).

Westwater, E. R., W. B. Sweezy, L. M. McMilllin, and C. Dean, "Determination of atmospheric temperature profiles from a statistical combination of ground-based profiler and operational NOAA 617 satellite retrievals". J. Climate Appl. Meteor., 23, 689-703. (1984).

Westwater, E. R., 1993: Ground-based microwave remote sensing of meteorological variables. Atmospheric Remote Sensing by Microwave Radiometry, M. A. Janssen, Ed., John Wiley & Sons, 145-213. (1993).

Westwater, E. R., "Remote sensing of tropospheric temperature and moisture by integrated observing systems," Bull. Amer. Meteor. Soc.,78, 1991-2006 (1997), (AMS Remote Sensing Lecture).

Westwater, E. R.,Yong Han, and Fred Solheim, "Resolution and accuracy of a multi-frequency scanning radiometer for temperature profiling", Microwave Radiometry and Remote Sensing of the Earth's Surface and Atmosphere, P. Pampaloni and S. Paloscia, Eds. VSP Press, 129-135, (2000).

Bayesian Techniques in Remote Sensing

NAZZARENO PIERDICCA
Dept. Electronic Engineering, University "La Sapienza" of Rome, Rome – Italy

Key words: Bayes, inversion methods, remote sensing, estimation

Abstract: Hereafter we give some basic concepts about decision theory and parameter estimation that are very often used in remote sensing problems. In particular, Bayesian criterion to test hypothesis is described to solve a data classification problem. Bayesian estimation of unknown parameters is presented as a method to invert direct electromagnetic models. The use of contextual information when dealing with images is also shortly described.

1. INTRODUCTION

Hereafter we give some basic concepts about decision theory and parameter estimation that are very often used in remote sensing problems. The subject is not exhaustively treated and for a more complete formulation the reader is referred to the textbooks in the bibliography. However, what follows should be enough to understand specific remote sensing applications of Bayesian estimation that are reported in the referenced papers.

The retrieval of information on the remotely sensed targets (i.e., atmosphere, ocean, land) is basically an ***inverse problem***. Most of the time it is an ill-posed problem, since many geophysical parameters concur to determine the quantities detected by the sensor (the observations) and different combinations of those parameters can give rise to similar observed quantities. Moreover, errors are always present in the measurements as well as in the knowledge of the causal model (the so called ***direct model***) that relates the parameters to the observations. The Bayesian approach offers a rigorous probabilistic framework to face the problem. In fact, it is able to incorporate the direct model as well as the a priori information.

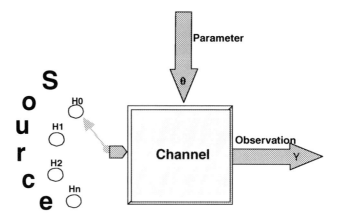

Figure 1. *Classification* is a hypothesis testing problem or decision problem, i.e., decide H_0, H_1.... from the observed **y**. For each hypothesis, the observation can be associated to different unknown parameters **θ**.

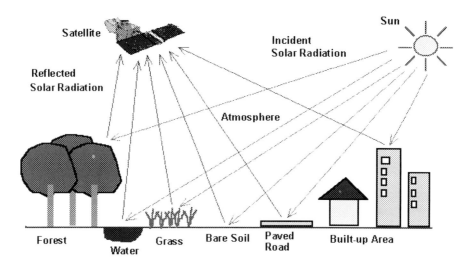

Figure 2. Retrieval of geophysical parameters is a parameter estimation problem, i.e., once H_i is assumed, estimate **θ** from **y**.

These may be available in probabilistic form, both on target properties and measurement/model errors, thus providing additional constraints to overcome the ill-poseness of the inverse problem.

Bayesian techniques in remote sensing

The Bayesian techniques have been greatly expanded in the field of radar and communication technologies, that is applications where it is needed to face a decision problem or a parameter estimation problem.

Let us assume a source is emitting several possible outputs H_0, H_1.... and that each of them, being transferred throughout a system (e.g., a communication channel), determines an observation **y** at the reception point. The hypothesis testing problem is to decide H_0, H_1.... from the observed **y**, as shown in figure 1. Likewise in remote sensing, based on the observed value of some radiative quantities, the classification rule decides as to which class the observed target belongs to. In addition, for each class, the observation is influenced by several parameters, depending on the target, the sensor, the interposed atmosphere and different sources of errors (noise). Therefore, the retrieval of geophysical parameters from remotely sensed data is a parameter estimation problem, i.e., once Hi is assumed, estimate a number of parameters (θ_1, θ_2 ....) associated with the observed **y**. This is depicted in figure 2.

2. A SURVEY OF PROBABILITY THEORY

In this section we remind some basic concepts of probability theory in order to introduce the notation that will be used in the following sections. Capital letter E indicates an event and its probability is indicated by P(E), being $0 \leq P(E) \leq 1$; P=0 in case of an impossible event. A random variable is indicated with a capital letter X and a value it can assume (a realization) in the interval [a,b] is represented by a lower-case letter x. A discrete random variable (whose values x are finite or countable) and a continuous random variable are indicated in the same way. For a continuous random variable, the probability density function (p.d.f.) $p_X(x)$ is such that $p_X(x)dx$ equals the probability of X to assume values between x and x+dx. It follows:

$$\int p_X(x)dx = 1 \tag{1}$$

A multidimensional random variable (or a random vector) is indicated by a bold capital letter **X** and its p.d.f. is therefore $p_\mathbf{X}(\mathbf{x})$.

The joint probability of events A and B is P(A and B), whilst for continuous random variables the joint p.d.f. is indicated by $p_{X,Y}(x,y)$. Similarly, the probability of a conditional event A|B (i.e., A conditioned to B) is P(A|B), whilst the conditional p.d.f. is represented by $p_{X|Y}(x|y)$.

The Bayes theorem states that P(A and B)=P(A|B)·P(B) or, for continuous variables $p_{X,Y}(x,y)=p_{X|Y}(x|y)p_Y(y)$. Statistically independent events are such that P(A|B)=P(A) and therefore it descends that P(A and

B)=P(A)·P(B), whilst for statistically independent random variables $p_{X|Y}(x|y)=p_X(x)$ and consequently $p_{X,Y}(x,y)=p_X(x)p_Y(y)$.

The expected value of a continuous random variable X is given by:

$$E[X] = m_X = \int x \cdot p_X(x)dx \qquad (2)$$

and the variance is the expectation of $(X-E[X])^2$, that is $\sigma^2_X = E[(X-E[X])^2]$; the correlation coefficient between two random variables X and Y is the expectation of $(X-E[X])(Y-E[Y])$ divided by the product of the standard deviations, that is $\rho_{XY}=E[(X-m_X)(Y-m_Y)]/\sigma_X \sigma_Y$.

These are particular cases of a more general one. Considering that a function of a random variable X is another random variable Y=g(X), its expectation is given by:

$$E[g(X)] = \int g(x) \cdot p_X(x)dx \qquad (3)$$

For a discrete random variable we can immediately extend the same formulas by considering the p.d.f. as a serious of Dirac functions centered at the values the discrete variable can assume and whose integrals equal the probability of that value. The expectation in equation (3) reduces to the summation of each probability multiplied by the appropriate function of the random variable.

3. BAYESIAN CRITERION FOR TESTING HYPOTHESIS

Let us assume that the probabilities of occurrence of the source outputs $H_0, H_1 ... H_{M-1}$ i.e., the *a priori* probabilities $P(H_0), P(H_1)... P(H_{M-1})$ are known. The observation space Z ($\mathbf{y} \in Z$) is partitioned into subspaces Z_i ($Z=Z_0 \cup Z_1 \cup .... Z_{M-1}$) and we decide H_i is true (decision D_i) if $\mathbf{y} \in Z_i$. We have M^2 alternative i.e., decide D_i (i=0, 1,... M-1) when H_j (j=0, 1, ... M-1) is true. Since the consequence of a decision may be different from the consequence of another, a cost is assigned to each possible alternative C_{ij}, i,j=0,1...., M-1. The case i=j represents a correct decision, so that for the cost it is assumed $C_{ij} > C_{ii}$ $\forall i \neq j$. The goal is to minimize the risk R, that is defined as the expected value of the cost. Being the cost a discrete random variable, its expected value is given by the following summation, as explained at the end of section 2.:

Bayesian techniques in remote sensing 53

$$R = \sum_i \sum_j C_{ij} P(D_i, H_j) = \sum_i \sum_j C_{ij} P(D_i | H_j) P(H_j) = \qquad (4)$$

$$= \sum_i \sum_j P(H_j) C_{ij} \int_{Z_i} p_{Y|Hj}(\mathbf{y} | H_j) d\mathbf{y}$$

where $P(D_i, H_j)$ is the joint probability of H_j being true and making decision D_i, i.e., the probability of cost C_{ij}. Note that $P(D_i|H_i)$ is the probability of a correct decision when H_i is true, whilst $P(D_i|H_j) \; \forall i \neq j$ is the probability of an erroneous decision. They both equal the probability of $\mathbf{y}$ belonging to subspace Z_i once the source output H_j is given ($\forall j$). This justifies the last equality.

By defining the ***Likelihood ratio*** $\Lambda_i(y)$, i=1,2, ... M-1 as:

$$\Lambda_i(\mathbf{y}) = \frac{p_{Y|Hi}(\mathbf{y} | H_i)}{p_{Y|H0}(\mathbf{y} | H_0)} \qquad i = 1, 2, ..., M-1 \qquad (5)$$

it can be demonstrated [Barkat, 1991] that, in order to minimize the risk, the decision rule should choose the hypothesis for which the following minimum is found with respect to i:

$$\min_i \{J_i(y)\} = \min_i \left\{ \sum_{j \neq i} P(H_j)(C_{ij} - C_{jj}) \Lambda_j(y) \right\} \qquad (6)$$

For example, in case of a binary decision between H_0 or H_1, we have to compare $J_0(y)$ against $J_1(y)$. Since each of them reduce to only one term of the summation and $\Lambda_0(y)=1$, comparing $J_0(y)$ against $J_1(y)$ corresponds to test the Likelihood ratio $\Lambda(y)=\Lambda_1(y)$ against a threshold η as in the following:

$$\Lambda(\mathbf{y}) = \frac{p_{Y|H_1}(\mathbf{y} | H_1)}{p_{Y|H_0}(\mathbf{y} | H_0)} \begin{array}{c} H_1 \\ > \\ < \\ H_0 \end{array} \frac{P(H_0)(C_{10} - C_{00})}{P(H_1)(C_{01} - C_{11})} = \eta \qquad (7)$$

Where the notation means that we choose H_1 if $\Lambda(y)$ is greater then the threshold, whilst we choose H_0 in the opposite case.

3.1 The Maximum A Posterior Probability criterion

If one wants to minimize the error probability, costs for a correct decision should be posed to zero whilst costs of any wrong decision should equal one, i.e., Cii=0 and Cij=1 $\forall i \neq j$. In this event, by substituting in equation (6), we obtain:

$$\min_i\{J_i(y)\} = \min_i\left\{\sum_{\substack{j=0\\j\neq i}}^{M-1} P(H_j)\Lambda_j(y)\right\} = \min_i\left\{\sum_{\substack{j=0\\j\neq i}}^{M-1} P(H_j)\frac{p_{Y|Hj}(y|H_j)}{p_{Y|H0}(y|H_0)}\right\} \quad (8)$$

by using $P(A|B)P(B) = P(B|A)P(A)$ (the Bayes theorem) and considering that the addition of all the probabilities $P(H_i|y)$ gives one, it is found:

$$\min_i\left\{\sum_{\substack{j=0\\j\neq i}}^{M-1} P(H_j)\frac{p_{Y|Hj}(y|H_j)}{p_{Y|H0}(y|H_0)}\right\} = \min_i\left\{\sum_{\substack{j=0\\j\neq i}}^{M-1} P(H_j|y)p_Y(y)\right\} =$$

$$= \min_i\{p_Y(y)[1 - P(H_i|y)]\}$$
(9)

Therefore, minimizing the error probability leads to the so called **Maximum A Posterior Probability** (MAP) criterion. In fact, the decision rule becomes searching for the hypothesis H_i that maximizes the conditional posterior probability $P(H_i|y)$ of hypothesis H_i given the observation y. In real problems $P(H_i|y)$ is unknown. However, it can be usually obtained from the conditional probability of observation y given H_i and the prior probability $P(H_i)$, so that the MAP criterion requires to search for a maximum with respect to i, as in the following:

$$\max_i\{P(H_j|y)\} = \max_i\left\{\frac{p_{Y|Hj}(y|H_j)P(H_j)}{p_Y(y)}\right\} = \max_i\{p_{Y|Hj}(y|H_j)P(H_j)\} \quad (10)$$

Where $p_Y(y)$ is discarded since, once the observation is given, it does not participate to the maximization process. If we do not know the prior probability we assume $P(H_i)$=constant and we maximize $p_{Y|Hi}(y|H_i)$. In this particular case we search for the maximum of the Likelhood ratio, that is we obtain the so called **Maximum Likelhood** (ML) criterion:

$$\max_i\{p_{Y|Hj}(y|H_j)\} \quad (11)$$

4. BAYESIAN CRITERION FOR ESTIMATING A RANDOM PARAMETER

Let us assume we have come to a decision in favor of the true hypothesis H_i, but the observation is still dependent on unknown parameters θ_1, θ_2, ... We wish to estimate the value of θ from a finite number of samples of the

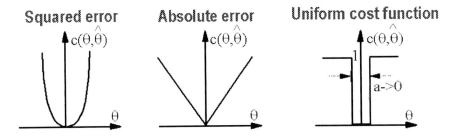

Figure 3: cost function for a scalar parameter. Squared error (left panel), absolute error (central panel) and uniform cost (right panel) correspond to MMSE, Median and MAP criteria, respectively.

observation **y**, i.e. predict θ. The function (a statistic) $\hat{\Theta} = g(\mathbf{Y})$ we use for estimating θ is a random variable called the *estimator* of θ. The specific value $\hat{\theta} = g(\mathbf{y})$ is called the *estimate* of θ.

Let us assume the parameter is a random variable Θ and the *prior* probability $p_\Theta(\theta)$ is available. A cost $c(\theta, \hat{\theta})$ is assigned to each pair of estimate $\hat{\theta}$ and real value θ of the parameter. The goal is to minimize the expected value of the cost, that is called the risk R, and is given by:

$$R = E[c(\theta,\hat{\theta})] = \iint c(\theta,\hat{\theta}) \, p_{\Theta,\mathbf{Y}}(\theta,\mathbf{y}) d\theta d\mathbf{y} \qquad (12)$$

Since only error $\hat{\theta} - \theta$ is relevant, three cases are considered, as depicted in figure 3 for a scalar parameter. The cost function is assumed either the square of the error, its absolute value or a stepwise function equal to zero for small errors and equal to one elsewhere.

4.1 Minimum Mean Square Estimate and median of $p_{\Theta|\mathbf{Y}}(\theta|\mathbf{y})$

The estimator which minimizes the risk for a squared error is called the **Minimum Mean Square Estimator**. The corresponding risk is given by:

$$R_{MMSE} = \iint (\theta - \hat{\theta}_{MMSE})^2 p_{\Theta,\mathbf{Y}}(\theta,\mathbf{y}) d\theta \, d\mathbf{y} \qquad (13)$$

Using Bayes theorem $p_{\Theta,\mathbf{Y}}(\theta,\mathbf{y}) = p_{\Theta|\mathbf{Y}}(\theta|\mathbf{y}) \, p_\mathbf{Y}(\mathbf{y})$ and the risk can be rewritten as:

$$R_{MMSE} = \int p_\mathbf{Y}(\mathbf{y}) d\mathbf{y} \left[\int (\theta - \hat{\theta}_{MMSE})^2 p_{\Theta|\mathbf{Y}}(\theta|\mathbf{y}) d\theta \right] \qquad (14)$$

Since $p_Y(y)$ is non-negative, minimizing R_{MMSE} is equivalent to minimize the quantity in bracket. It can be demonstrated (Barkat, 1991) we can obtain the following MMSE estimator:

$$\hat{\theta}_{MMSE} = E[\theta|Y] = \int \theta\, p_{\Theta|Y}(\theta|y)d\theta \qquad (15)$$

Therefore the Minimum Mean Square Estimate (MMSE) is the conditional mean of θ given Y. From equation (14) and (15), it comes out that it minimizes, in the average, the conditional variance of θ given Y var$[\theta|Y]$ (i.e., it also implies a *Minimum Variance of Error*).

As for the estimator minimizing the absolute error, the risk is given by:

$$R_{abs\_err} = \iint |\theta - \hat{\theta}_{abs\_err}|\, p_{\Theta,Y}(\theta,y)d\theta\, dy \qquad (16)$$

It can be demonstrated that the corresponding estimate $\hat{\theta}_{abs\_err}$ satisfies the following equation (Barkat, 1991):

$$\int_{-\infty}^{\hat{\theta}_{abs\_err}} p_{\Theta|Y}(\theta|y)d\theta = \int_{\hat{\theta}_{abs\_err}}^{+\infty} p_{\Theta|Y}(\theta|y)d\theta \qquad (17)$$

That means $\hat{\theta}_{abs\_err}$ is the median of the conditional p.d.f. of θ given Y.

4.2 Maximum A Posteriori probability estimate

By considering the uniform cost function, the corresponding risk is given by:

$$R_{MAP} = \int dy \left[\int_{-\infty}^{\hat{\theta}_{MAP}-a/2} p_{\Theta,Y}(\theta,y)d\theta + \int_{\hat{\theta}_{MAP}+a/2}^{+\infty} p_{\Theta,Y}(\theta,y)d\theta \right] \qquad (18)$$

By applying the Bayes theorem $p_{\Theta,Y}(\theta,y) = p_{\Theta|Y}(\theta|y)\, p_Y(y)$ and observing that the unlimited integral of $p_{\Theta|Y}(\theta|y)$ should be one, it turns out:

$$R_{MAP} = \int p_Y(y)dy \left[1 - \int_{\hat{\theta}_{MAP}-a/2}^{\hat{\theta}_{MAP}+a/2} p_{\Theta|Y}(\theta|y)d\theta \right] \qquad (19)$$

Therefore the decision rule reduces to search for θ that maximizes the conditioned posterior probability $p_{\Theta|Y}(\theta|y)$ of parameter θ given the observation y. For this reason it is called the **Maximum A Posterior Probability** (MAP) criterion. By using again the Bayes theorem and the propriety of the logarithm being a monothonic function, MAP criterion can be stated in this way:

Bayesian techniques in remote sensing

$$\max_{\theta}\{p_{\Theta|Y}(\theta|y)\} = \max_{\theta}\{p_{Y|\Theta}(y|\theta)p_{\Theta}(\theta)\} \Rightarrow \frac{\partial ln[p_{\Theta|Y}(\theta|y)]}{\partial \theta_i} = 0 \quad (20)$$

where, being $p_{\Theta|Y}(\theta|y)$ difficult to determine, it has been expressed by the conditional probability of observation y given θ and the prior probability $p_\Theta(\theta)$ of the parameter. Note that for $p_{\Theta|Y}(\theta|y)$ unimodal and symmetric (e.g., Gaussian) MMSE, median and MAP criteria coincide.

4.2.1 Maximum Likelihood Estimator as a particular case of MAP

Now let us assume parameter θ is a constant unknown quantity. Since the parameter influences the observation, we can only state that the p.d.f. of observation Y is function of the parameter θ, i.e., $p_Y(y) = p_Y(y;\theta)$.

Using the previous treatment for the random parameter case, $p_Y(y;\theta)$ can be now identified with the p.d.f. of Y given the parameter Θ i.e., $p_{Y|\Theta}(y|\theta)$, but this time no probabilistic information are available on Θ. The MAP criterion requires to find the maximum of equation (20). Being the prior probability of θ unknown, it is required to assume $p_\Theta(\theta)$ is constant and therefore to maximize $p_{Y|\Theta}(y|\theta)$. In this case we obtain the **Maximum Likelhood Estimate** (MLE), i.e., the value $\hat{\theta}$ which, assigned the observed y, maximizes the likelihood function $\Lambda(\theta)=p_Y(y;\theta)$ or the log-likelihood function $\ln[\Lambda(\theta)]$:

$$\max_{\theta}\{\Lambda(\theta)\} = \max_{\theta}\{p_Y(y;\theta)\} \Rightarrow \max_{\theta}\{p_Y(y|\theta)\} \Rightarrow \frac{\partial}{\partial \theta_i} ln[p_Y(y;\theta)] = 0 \quad (21)$$

Note that the maximum likelihood criterion to estimate non random parameters has been derived as a particular case of MAP.

4.3 Example: unknown constant or random quantity with noise

Let us consider a radar whose observed output y contains either noise with zero mean and σ standard deviation (H_0 hypothesis) or an unknown echo to which an additive noise is superimposed (H_1 hypothesis). Consider three different problems as in the following:
A. Determine the test to decide H_0 or H_1 from the observed y.
B. In case of H_1 and echo is an unknown constant m, estimate it from K independent observations of y.
C. In case echo is a random variable θ with mean m and standard deviation σ, estimate it from K independent observations of y.
Hereafter you will find the solution to the three problems.

A. The p.d.f. of y in case of H_0 or H_1 is Gaussian with mean equal to 0 or m, respectively. Then the test on likelihood function assumes the following form:

$$\Lambda(y) = \frac{p_{Y|H_1}(y|H_1)}{p_{Y|H_0}(y|H_0)} = \frac{\frac{1}{\sqrt{2\pi}\sigma}e^{-\frac{(y-m)^2}{2\sigma^2}}}{\frac{1}{\sqrt{2\pi}\sigma}e^{-\frac{y^2}{2\sigma^2}}} = e^{-\frac{m^2-2ym}{2\sigma^2}} \underset{H_0}{\overset{H_1}{\gtrless}} \eta \qquad (22)$$

By computing the logarithm of both members, the test on the log-likelihood function allows one to decide between the two hypothesis:

$$\ln\Lambda(y) = \frac{m}{\sigma^2}y - \frac{m^2}{2\sigma^2} \underset{H_0}{\overset{H_1}{\gtrless}} \ln\eta \qquad (23)$$

B. Considering K independent samples of the radar output, the p.d.f. of the joint random variable **Y** is the product of K Gaussian with mean m and standard deviation σ. The parameter to estimate is an unknown constant and therefore we apply the ML criterion. It consists of the maximization of a log-likelihood function as in the following:

$$\frac{\partial \ln[p_Y(y;m)]}{\partial m} = \sum_k \frac{y_k - m}{\sigma^2} = 0 \Rightarrow m = \frac{1}{K}\sum_k y_k \qquad (24)$$

from which it comes out the following estimate of the unknown constant:

$$\hat{m}_{ML} = \frac{1}{K}\sum_k Y_k \qquad (25)$$

C. In this case we know the p.d.f. of the unknown parameter. We apply the MAP criterion that requires the minimization of the posterior p.d.f. $p_{\Theta|Y}(\theta|y)$ of θ given y, i.e.,

$$\max_\theta \{p_{\Theta|Y}(\theta|y)\} = \max_\theta \left\{\frac{p_{Y|\Theta}(y|\theta)p_\Theta(\theta)}{p_Y(y)}\right\} = \max_\theta \left\{\prod_k \frac{1}{\sqrt{2\pi}\sigma}e^{-\frac{(y_k-m)^2}{2\sigma^2}} \frac{1}{\sqrt{2\pi}\sigma}e^{-\frac{m^2}{2\sigma^2}}\right\} \qquad (26)$$

from which, using the logarithm, we found the following estimate that is slightly different from what we found in the previous point:

$$\hat{\theta}_{MAP} = \frac{1}{K+1}\sum_k Y_k \qquad (27)$$

4.4 Criteria of "goodness": Unbiased and Minimum Variance estimates

Given θ (the true value) a "good" estimator is a random variable which is expected to be:

- **Unbiased**: estimates should be centered around the true value θ. Once θ is fixed, the expected value of estimates should be:

$$E(\hat{\theta}|\theta) = \int \hat{\theta} p_{Y|\Theta}(y|\theta)dy = \theta \qquad (28)$$

- Unbiased with **minimum variance**: the "dispersion" of the estimates around the true value is represented by their variance. It should be minimum compared to all unbiased estimators:

$$\text{var}[\hat{\theta}|\theta] = \text{var}[(\hat{\theta}-\theta)|\theta] = \int(\theta-\hat{\theta})^2 p_{Y|\Theta}(y|\theta)dy = \min \qquad (29)$$

Being computation of $\text{var}[\hat{\theta}|\theta]$ very difficult (it depends on θ), the Cramer-Rao inequality can be useful to verify the "goodness" of the monodimensional estimator. It states (Barkat, 1991):

$$\text{var}(\hat{\theta}-\theta)|\theta) \geq 1 \Big/ E\left\{\left[\frac{\partial \ln p_{Y|\Theta}(y|\theta)}{\partial \theta}\right]^2\right\} \qquad (30)$$

5. APPLICATIONS

5.1 Classification of RS images

The scope of image classification is to assign each pixel to a class i (e.g., land use or land cover classes) according to a legend. Given the observation vector **y** (of dimension D), we apply the MAP criterion, that is we maximize $p_{I|Y}(i|y)$. By considering that the logarithmic function increases monotonically, we can express the MAP criterion in the following way:

$$\max_i\{p_{I|Y}(i|y)\} = \max_i\{p_{Y|i}(y|i)P(i)\} = \max_i\{\ln[p_{Y|i}(y|i)] + \ln[P(i)]\} \qquad (31)$$

Let us assume the observations **y** within class i vary from pixel to pixel because of many random uncorrelated sources of variations. We can assume **y** conditioned to class i is a Gaussian random variable with mean m_i and covariance matrix C_i:

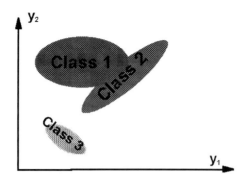

Figure 4: spectral signatures in two dimensions (two observations or channels) of three individual classes. Ellipses represent points of equal $p_{Y|I}(y|i)$.

$$p_{Y|I}(\mathbf{y}|i) = \frac{1}{(2\pi)^{\frac{D}{2}}\sqrt{\det(\mathbf{C}_i)}} \exp-[\frac{1}{2}(\mathbf{y}-\mathbf{m}_i)^T \mathbf{C}_i^{-1}(\mathbf{y}-\mathbf{m}_i)] \qquad (32)$$

Using the logarithm, it comes out that MAP requires to minimize the following discriminant function $d(\mathbf{y},i)$ with respect to i:

$$\min_i \{d(\mathbf{y},i)\} = \min_i \{(\mathbf{y}-\mathbf{m}_i)^T \mathbf{C}_i^{-1}(\mathbf{y}-\mathbf{m}_i) + \ln[\det(\mathbf{C}_i)] - 2\ln P(i)\} \qquad (33)$$

The **spectral signature** of class i, (i.e., m_i and C_i), is derived by a sufficient number of pixels whose class is known (a **training set**) by:

$$\mathbf{m}_i = E(X) \approx \frac{1}{N}\sum_{j=1}^{N}\mathbf{x}_j \quad \mathbf{C}_i = E[(\mathbf{x}-\mathbf{m})^T(\mathbf{x}-\mathbf{m})] \approx \frac{1}{N-1}\sum_{j=1}^{N}(\mathbf{x}_j-\mathbf{m})^T(\mathbf{x}_j-\mathbf{m}) \qquad (34)$$

Note that $d(\mathbf{y},i)$, thus the classification rule, accounts for the distance of observed **y** from the class center, the spreading of observations within the class and the class prior probability, when available.

5.2 Inversion of a causal model in presence of noise

Suppose $g(\theta)$ relates the unknown parameters to the observed quantities. The unknown parameters represent some properties of the remotely sensed target whilst the observations are some radiative quantities measured by a sensor at different frequency, polarization, etc.. Therefore $g(\theta)$ represents the so called *direct model* in a remote sensing problem. Let us assume **n** is any source of departure of the observed **y** from $g(\theta)$. It can be due either to error in the direct model and to measurement errors.

Bayesian techniques in remote sensing

If it is assumed that error **n** is independent on **θ**, it is the case of an *additive*

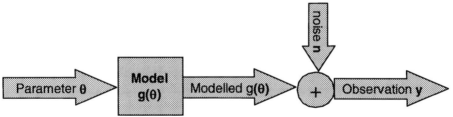

noise for which we write **y**=g(**θ**)+**n**. **n** is independent from **θ** and it is usually

Figure 5. Block diagram representing the acquisition of remote sensing data affected by an additive noise

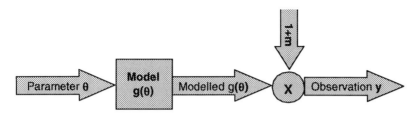

Figure 6. Block diagram representing the acquisition of remote sensing data affected by a multiplicative noise

assumed Gaussian with zero mean and covariance C_n. This is depicted in Figure 5.

When **n** is not independent from **θ**, a relevant case is represented by the equation **y**=g(**θ**)+**m**g(**θ**) with **m** independent on **θ** and E[**m**]=0. This is the particular case of a multiplicative model which is depicted in Figure 6.

5.2.1 Single observation in additive noise and Gaussian prior p.d.f.

The observations **y** for a given parameter **θ** fluctuate around g(**θ**) because of error **n**, therefore the p.d.f. of **y**|**θ** is given by $p_{Y|\Theta}(y|\theta) = p_{Y|\Theta}(y - g(\theta)) = p_N(n)$. The posterior probability can be derived by using the Bayes theorem:

$$p_{\Theta|Y}(\theta|y) = p_{Y|\Theta}(y|\theta) \cdot p_\Theta(\theta) / p_Y(y) = p_N(y - g(\theta)) \cdot p_\Theta(\theta) / p_Y(y) \quad (35)$$

Assume **θ** is Gaussian with mean m_θ and covariance C_θ and **n** is an additive Gaussian noise, with zero mean and covariance C_n. The MAP criterion requires to minimize the following discriminant function d(**y**,**θ**) with respect to **θ**.

$$\min_{\theta}\{d(\mathbf{y},\theta)\} = \min_{\theta}\{[\mathbf{y}-g(\theta)]^T \mathbf{C_n}^{-1}[\mathbf{y}-g(\theta)] + [\theta-\mathbf{m_\theta}]^T \mathbf{C_\theta}^{-1}[\theta-\mathbf{m_\theta}]\} \quad (36)$$

Note that the discriminant function balances the departure of the observation from model and of the estimate from the known mean values of the parameter. The two terms are properly weighted according to the error characteristics represented by $\mathbf{C_n}$. For uncorrelated noise components with equal variance σ_n^2, $\mathbf{C_n} = \sigma_n^2 \mathbf{I}$ and therefore the MAP criterion becomes:

$$\min_{\theta}\{d(\mathbf{y},\theta)\} = \min_{\theta}\left\{\frac{1}{\sigma_n^2}\sum_{l=1}^{D}[y_l - g_l(\theta)]^2 + [\theta-\mathbf{m_\theta}]^T \mathbf{C_\theta}^{-1}[\theta-\mathbf{m_\theta}]\right\} \quad (37)$$

5.2.2 Additive noise special cases: linear direct model

In case of a linear direct model $g(\theta) = \mathbf{K}\theta$, where $\mathbf{K}$ is a matrix whose dimensions equal the number of parameters by the number of observations (channels). Assuming a Gaussian unknown parameter and an additive noise, $p_{\Theta|Y}(\theta|\mathbf{y})$ given by equation (35) is a symmetric and unimodal function of θ. Therefore MAP, MMSE and median estimators yield to the minimization of the following quadratic form:

$$d(\mathbf{y},\theta) = [\mathbf{y}-\mathbf{K}\theta]^T \mathbf{C_n}^{-1}[\mathbf{y}-\mathbf{K}\theta] + [\theta-\mathbf{m_\theta}]^T \mathbf{C_\theta}^{-1}[\theta-\mathbf{m_\theta}] \quad (38)$$

From which it can be found the well known solution of a multivariate linear regression problem:

$$\hat{\theta} = \mathbf{m_\theta} + \mathbf{C_\theta}\mathbf{K}^T(\mathbf{K}\mathbf{C_\theta}\mathbf{K}^T + \mathbf{C_n})^{-1}(\mathbf{y}-\mathbf{K}\mathbf{m_\theta}) \quad (39)$$

If the different components of noise and parameter are uncorrelated each other with equal variance, i.e., $\mathbf{C_n} = \sigma_n^2 \mathbf{I}$ and $\mathbf{C_\theta} = \sigma_\theta^2 \mathbf{I}$, the discriminant function is expressed by a quadratic form plus a regularization term proportional to the noise variance and the quantity to be minimized becomes:

$$d(\mathbf{y},\theta) = [\mathbf{y}-\mathbf{K}\theta]^T[\mathbf{y}-\mathbf{K}\theta] + \sigma_n^2/\sigma_\theta^2 [\theta-\mathbf{m_\theta}]^T[\theta-\mathbf{m_\theta}] \quad (40)$$

Finally, if no information are available on noise and on prior probability of the parameter, the ML criterion only minimizes the departure of the observation from the model. This is the simple least square criterion, which is not protected against the effects of the errors:

$$d(\mathbf{y},\theta) = [\mathbf{y}-\mathbf{K}\theta]^T[\mathbf{y}-\mathbf{K}\theta] \quad (41)$$

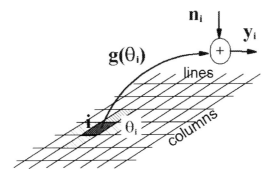

Figure 7: conceptual scheme for a classification problem that accounts for contextual information

5.3 The contextual information in RS images

Many remotely sensed data has a lattice nature (image), each observation y_i being taken at site (pixel) i, with parameter θ_i. We indicate with $[\theta]$ and $[y]$ the sets of all parameters and related observations in the scene. $\theta_{s/i}$ is the set of parameters of those pixels (neighborhood) from which θ_i is not statistically independent. $p(\theta_i|\theta_{s/i})$ is the conditional probability of parameter in i given all neighbors.

To apply the MAP criterion we have to maximize the posterior probability with respect to θ_i. By applying the Bayes theorem and discarding terms that do not depend on θ_i, the MAP criterion to estimate parameters of all pixels in the image, given all the observations, reduce to:

$$\max_{\theta_i}\{p([\theta]|[y])\} = \max_{\theta_i}\{p(\theta_i,[\theta]_{j\neq i}|y_i,[y]_{j\neq i})\} = \max_{\theta_i}\{p(\theta_i|[\theta]_{j\neq i},y_i,[y]_{j\neq i})\} \quad (42)$$

There is a causal relation between θ_i and y_i (i.e., θ_i does not affect all the observations $[y]_{i\neq j}$) and θ_i depends only on subset $\theta_{s/i}$ of neighbor's parameters. Therefore:

$$\max_{\theta_i}\{p(\theta_i|[\theta]_{j\neq i},y_i,[y]_{j\neq i})\} = \max_{\theta_i}\{p(\theta_i|\theta_{s/i},y_i)\} \quad (43)$$

By applying Bayes theorem again and discarding terms that do not depend on θ_i it is obtained:

$$\max_{\theta_i}\{p(\theta_i|\theta_{s/i},y_i)\} = \max_{\theta_i}\{p(\theta_i|y_i)p(\theta_{s/i}|\theta_i)p_{\Theta_i}(\theta_i)\} \quad (44)$$

and the estimation rule that accounts for the contextual information becomes:

$$\max_{\theta_i}\{p(\theta_i \mid y_i)p(\theta_i \mid \boldsymbol{\theta}_{s/i})\} \tag{45}$$

Besides the posterior probability at each pixel, it requires the conditional probability of the parameter at pixel i given all its neighbors (i.e., the contextual information). Same considerations apply to develop a classification rule that accounts for the context.

6. REFERENCES

Barkat M., "Signal Detection & Estimation", Artech House Ed., 1991

Datcu M., K. Seidel, M. Walessa, "Spatial Information Retrieval from Remote-Sensing Images - Part I: Infromation Theorethical Perspective", IEEE Trans. Geosci. Rem. Sens., 36, 4, September 1998

Davis T., Z. Chen, J-N Hwang, L. Tsang, E. Njoku, "Solving Inverse Problems by Bayesian Iterative Inversion of a Forward Model with Application to Parameter Mapping Using SMMR Remote Sensing Data", IEEE Trans. Geosci. Rem. Sens., 33, 5, September 1995

Evans K.F., J. Turk, T. Wong, and G. Stephens, "A Bayesian approach to microwave precipitation profile retrieval", J. Appl. Meteorol., 34, 1995.

Keihm S. J., K. A.Marsh, "New model-based Bayesian inversion algorithm for the retrieval of wet troposphere path delay from radiometric measurements", RadioScience, 33, 2, March-April 1998

Marzano F. S., A. Mugnai, G. Panegrossi, N. Pierdicca, E. A. Smith, J. Turk, "Bayesian Estimation of Precipitating Cloud Parameters from Combined Measurements of Spaceborne Microwave Radiometer and Radar", IEEE Trans. Geosci. Rem. Sens., 37, 1, January 1999

Papoulis, "Probability, Random Variables and Stochastic Process", McGraw-Hill, 1991

Pierdicca N., P. Basili, P. Ciotti, F. S. Marzano, "Inversion of electromagnetic models for estimating bare soil parameters from radar multifrequency and multipolarization data", Proc. of SPIE SAR Image Analysis, Modeling, and Techniques, Edited by F. Posa, Barcelona, Spain, 23-24 September 1998.

Pierdicca N., F.S. Marzano, G. d'Auria, Basili P., P. Ciotti, and A. Mugnai, "Precipitation retrieval from spaceborne microwave radiometers using maximum a posteriori probability estimation", IEEE Trans. Geosci. and Rem. Sens., 34, 1996

Richards J. A., "Remote Sensing Digital Image Analysis. An Introduction", Springer-Verlag Ed., 1986

Rieder M.J., G. Kirchengast, "An inversion algorithm for nonlinear retrieval problems extending Bayesian optimal estimation", Radio Science, 35, 1, January-February 2000.

Rogers, "Retrieval of Atmospheric Temperature and Composition From Remote Measurements of Thermal Radiation", Reviews of Geoph. and Space Physics, 4, November 1986

Solheim, J. R. Godwin, E. R. Westwater, Y. Han, S. J. Keihm, K. Marsh, R. Ware, "Radiometric profiling of temperature, water vapor and cloud liquid water using various inversion methods", Radio Science, 33, 2, March-April 1998

Chapter 2

Atmospheric Remote Sensing by Microwave and Visible-Infrared Sensors

Atmospheric Temperature Analysis Using Microwave Radiometry

STEPHEN J. ENGLISH
Satellite Radiance Assimilation Group - Met Office, United Kingdom

Key words: estimation

Abstract: Bayesian criterion to test hypothesis is described to solve a data classification problem. Bayesian estimation of unknown parameters is presented as a method to invert direct electromagnetic models. The use of contextual information when dealing with images is also shortly described

1.   INTRODUCTION

In order to obtain remotely sensed information about the temperature of the atmosphere using microwave radiometry it is necessary to sample the electromagnetic spectrum in the vicinity of the spectral lines which result from the magnetic dipole on the oxygen molecule. This magnetic dipole gives rise to spectral lines between 50 and 70 GHz and at 118 GHz. It is advantageous to use the lower frequency lines because the atmosphere is more transparent to cloud, and it is usually possible to achieve lower instrument noise. As a result instruments designed to provide temperature information have used the low frequency side of the complex of oxygen spectral lines between 50 and 57 GHz, as listed in Table 1.

Instrument	Launch year	Frequencies	Nadir field of view
NEMS	1972	53.6, 54.9, 58.8	200 km
SCAMS	1975	52.8, 53.8, 54.4	150 km
SSM/T	1978	50.5, 53.2, 54.3, 54.9, 58.4, 58.8, 59.0	175 km
MSU	1978	50.3, 53.7, 55.0, 57.9	110 km
AMSU	1998	12 channels 50.3-57.293	48 km

Table 1: The main nadir microwave radiometers for temperature sounding 1972-1998

In terms of information NEMS, SCAMS and MSU[*] were similar radiometers, although the latter had a smaller field of view size. The recently launched Advanced MSU has substantially more channels although the information these channels provide is not completely independent. The Special Sensor Microwave Temperature Sounder (SSM/T) provides information which is intermediate between that provided by AMSU and that provided by MSU, SCAMS and NEMS. The AMSU also has very low instrument noise and a smaller instantaneous field of view. AMSU is described in more detail by Saunders (1993).

The primary purpose of the microwave temperature sounding instruments is to provide information on atmospheric temperature in cloudy areas. Absorption by liquid water is not negligible at 50 GHz and the effects of precipitation and ice cloud can also not be ignored. However thin cirrus that would be opaque in the infra-red is completely transparent at 50 GHz. Thus most high cloud is transparent allowing very extensive use of channels which are sensitive to temperature changes in the upper troposphere. Low altitude clouds which contain high amounts of liquid water and deep frontal or convective clouds need to be detected and either the cloud contribution to the radiances modelled or these radiances excluded from the temperature analysis process. Typically 90-95% of channel radiances whose level of peak sensitivity is above 500 hPa can be safely used in the temperature analysis without modelling cloud. This falls to 65% for channels senstive to lower tropospheric temperature but this still compares favourably to the 20-30% of infra-red soundings which can be treated as cloud-free.

2. RADIATIVE TRANSFER AT 50-57 GHZ

Figure 1 shows how the shape of the spectral lines changes with altitude. At high altitude (low pressure) the individual spectral lines are well separated.

---

[*] MSU=Microwave Sounding Unit; NEMS=Nimbus-E Microwave Spectrometer; SCAMS=Scanning Microwave Spectrometer; SSM/T=Special Sensor Microwave Temperature Sounder; AMSU=Advanced MSU.

Atmospheric temperature analysis 69

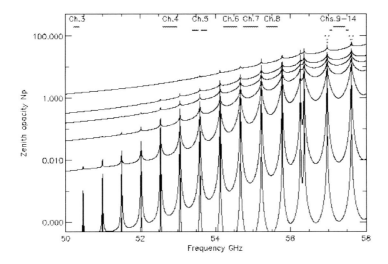

Figure 1: Oxygen absorption 50-58 GHz showing position of the AMSU channels. The seven curves show the total atmospheric opacity at 1000 hPa, 700 hPa, 500 hPa, 300 hPa, 100 hPa, 10 hPa and 1 hPa (highest opacity at highest pressure).

At lower altitude (higher pressure) the lines broaden until at 1000 hPa the individual spectral lines overlap to form an almost continuous absorption band.

Rosenkranz (1975, 1992) discusses the processes determining oxygen absorption at these wavelengths and how they are modelled. How strong the spectral absorption is determines the altitude at which a given radiance measurement has largest sensitivity. The change in radiance per unit change in temperature can be plotted against height to visualise the region of sensitivity for each channel. These curves are known as weighting functions.

The weighting functions for AMSU are shown in Figure 2. The weighting functions are broad (several kilometres deep) and overlap between channels. Therefore the radiances can only resolve deep features or determine the presence of a large but shallow feature without being able to determine its height. As the channel weighting functions overlap there are fewer independent pieces of information about temperature than there are radiance measurements.

For frequencies below 54 GHz the zenith opacity is low enough (<10) for a significant surface contribution to the measured top of atmosphere radiances. Therefore it is important to model the surface contribution if atmospheric temperature information is to be obtained from measurements between 50 and 54 GHz. AMSU channels 3, 4 and 5 (50.3, 52.8 and 53.596) are all below 54 GHz.

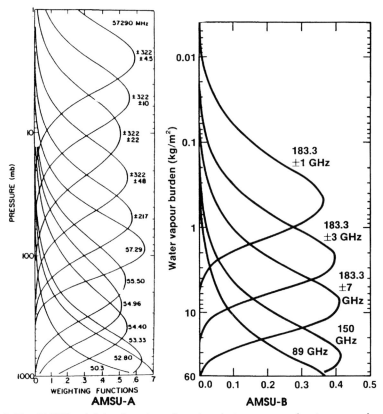

Figure 2: The AMSU weighting functions plotted against pressure and water vapour burden.

Modelling the surface contribution requires an accurate means of estimating or modelling the surface emissivity. For the ocean surface the emissivity can be modelled using a surface windspeed and surface skin temperature to within 1-2% using a geometric optics approach (Petty and Katsaros 1988). For land surfaces the emissivity depends on snow cover, surface wetness, vegetation and soil type. Variations in surface topography can also be important. English and Hewison (1998) discuss the representation of land surface microwave emissivity.

The contribution of liquid water emission to the measured radiances is also important. Again for frequencies above 54 GHz the contribution is likely to be very small, unless the liquid water is at high altitude (> 5km) (English et al. 2000). As the occurrence of sufficiently high values of liquid water is likely to be rare it is possible simply to detect these cases and only use field of views with low liquid water content for temperature analysis. Alternatively the cloud emission can be modelled and analysed simultaneously with the atmospheric temperature.

3. DATA INVERSION

The temperature analysis problem, or inverse problem, is one where we want to analyse a state vector **x** given some observations **y**. Ideally the state vector should contain all variables to which the observations **y** are sensitive. If we write the forward problem as **y**=**H**(**x**) then we can write the non-linear inverse problem using optimal estimation theory as,

$$\mathbf{x}_{n+1} = \mathbf{x}^b + \mathbf{W}_n \cdot (\mathbf{y} - \mathbf{y}_n - \mathbf{H}_n(\mathbf{x}^b - \mathbf{x}_n)) \tag{1}$$

where $\mathbf{x}^b$ is the background (e.g. from climatology or from a short range numerical weather forecast), $\mathbf{W}_n$ is the inverse operator calculated at the nth iteration, $\mathbf{x}_n$ and $\mathbf{x}_{n+1}$ are the nth and $n+1$th analysis of **x**, **y** is the observation vector (e.g. the 20 AMSU channel radiances), $\mathbf{y}_n$ is the nth estimate of y and H_n is the observation operator calculated at the nth iteration. This solution is optimal if $\mathbf{W}_n = \mathbf{B} \cdot \mathbf{H}_n^T (\mathbf{H}_n \cdot \mathbf{B} \cdot \mathbf{H}_n^T + \mathbf{O} + \mathbf{F})^{-1}$ in the sense that it minimises the total error variance (**B** is a matrix describing the background error covariance, **O** is a matrix describing the observation error covariance and **F** is a matrix describing the forward model error covariance). As written above it is a data assimilation problem as it is combining new observations with a background to create a new analysis. The equation makes no assumption about the number of dimensions of the problem so it could be applied to a single observation to produce a profile of temperature or it could be an instantaneous three dimensional analysis. The data analysis problem is discussed by Rodgers (1976, 1990) and applied specifically to TOVS data by Eyre *et al* (1993), Andersson *et al*. (1994), Gadd *et al*. (1995) and Derber and Wu (1998).

The analysis problem is shown figuratively in Figure 3. At the heart of the problem is an accurate forward model or observation operator. When very large quantities of data need to be processed in real time it is necessary to have a fast and accurate radiative transfer model. One such model is described by Saunders et al. (1999). Also fundamental is that quantities are compared in observation space i.e. we model the observations from our existing background and then interpret the differences. This ensures that the analysis is only modified from the background where there is information in the radiances and the radiances differ from those that are modelled from the background.

The analysis problem described by equation 1 and Figure 3 allows for radiances to be assimilated directly into a new analysis. In the late 1980s and early 1990s it was recognised that the information content of satellite radiance measurements by instruments like MSU and AMSU is rather low relative to the existing "knowledge" of the NWP system (Kelly et al. 1991).

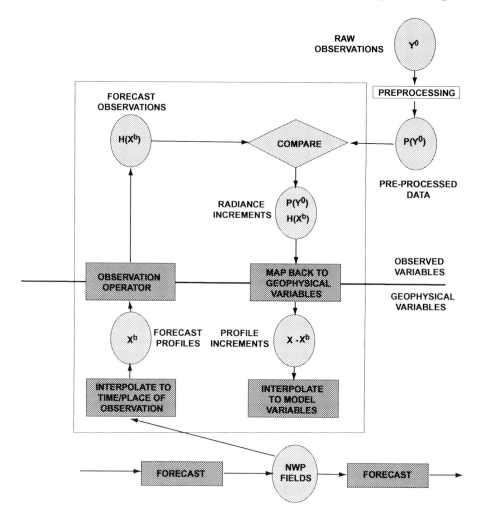

Figure 3. The data analysis problem (taken from Eyre 1997). Note that for this problem the observation operator is effectively the radiative transfer model since interpolation issues (which strictly are part of the observation operator) have been shown separately.

This is primarily because of the poor vertical resolution of the temperature profile provided by the radiances. Direct use of the radiances allows the radiance information to be assimilated with proper regard to their error characteristics and information content. As the accuracy of NWP systems has improved the need for more careful assimilation of radiances has also become clearer if positive impact is to be achieved. These issues are discussed further by Eyre and Lorenc (1989). Kelly (1999) and Renshaw (2000) have shown that directly assimilated radiances can now provide a clear improvement in forecast skill in both hemispheres.

4. ERROR SOURCES

Figure 4 shows the instrument noise variance for AMSU channels 5 to 11. These can be compared with the original specification for the AMSU-A instrument. The instrument noise variance achieved in space is about a quarter of that which was planned in the original specification. Thus AMSU is a considerably more accurate instrument than originally specified.

Despite this instrument noise remains the dominant term in the total error budget for AMSU channel 7 which peaks at 250 hPa and AMSU channel 6 which peaks at 400 hPa. This can be seen by comparing the instrument noise variance with the measured minus background variance in Figure 3. For AMSU channel 7 the instrument noise is more than 75% of the total observation minus model variance in the tropics. This contribution is only slightly lower when compared with global observation minus model variance.

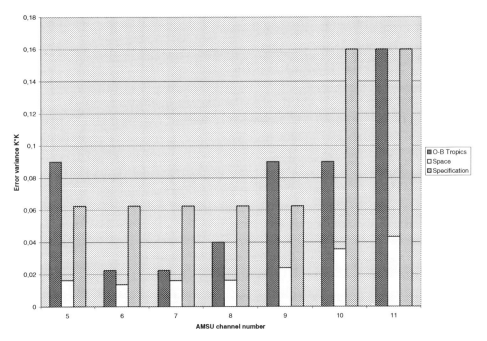

Figure 4: AMSU instrument noise variance in space compared to observation minus model differences in the tropics and the original AMSU-A specification.

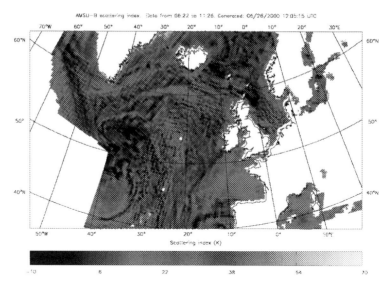

Figure 5: AMSU-B 89 GHz scattering index 26 June 2000 for the N. Atlantic

5. CLOUD DETECTION

Whilst less than 10% of AMSU observations are significantly affected by ice cloud and less than 40% water cloud it is nonetheless important to follow a policy of detecting and rejecting cloudy observations or of simultaneously analysing cloud and temperature. For the detection of ice cloud a scattering index approach is a popular technique (Grody and Ferraro 1992). In this technique the high frequency window channel(s) are compared with the low frequency window channels and if the difference is not consistent with our expectation in the absence of scattering then the difference may be assumed to be due to scattering. Figure 5 shows a typical image of 89 GHz scattering index for the North Atlantic Ocean.

Outside the tropics strong scattering at 89 GHz occurs in most passes but constitutes only a very small percentage of the data. Figure 5 shows a few cells of strong scattering for the AMSU-B 89 GHz channel on a cold weather front from 22W 54N to 32W 39N. The high values near coasts are caused by the different field of view size between AMSU-B and AMSU-A. The scattering index approach is used operationally at NESDIS (Grody et al. 1999) and the UK Met. Office (English et al. 1999) and is used as a precipitation product (e.g. Thoss et al. 2000).

The detection of liquid water is also important. The AMSU instrument has channels at 23.8 and 31 GHz specifically to identify liquid water. The differential absorption of water vapour and cloud liquid water makes this pair of channels very useful for providing information both on total integrated water vapour and total integrated liquid water. Over the sea it is straightforward to devise simple tests which flag areas of high liquid water content. Over the land it is not possible to detect liquid water with sufficient confidence using the window channels as the effects of the liquid water are often smaller for the window channels than the sounding channels. An example of a simple cloud liquid water detection scheme (English *et al.* 1997), based on AMSU channels 1, 2 and 3, is shown superimposed on AVHRR data in figure 6. The AMSU product highlights the areas of highest liquid water content. Different approaches to cloud detection for AMSU are discussed in English *et al.* (1999).

6. TEMPERATURE SOUNDING OVER LAND

Temperature sounding over land is more difficult than over the sea. Firstly there is the issue of cloud detection already discussed in section 5. As land surface emissivity is usually above 0.90 most land surfaces emit at a temperature similar to the atmosphere. This makes it difficult to distinguish atmospheric and surface emission. For some surfaces the emissivity is not well known. Also the land surface is more heterogeneous than the ocean, has a less well known skin temperature and has changes in altitude. All these effects need to be fully taken into account when analysing temperature over land. Inevitably this results either in higher forward model errors (e.g. because emissivity can not be computed) or fewer channels used (because we can't accurately model them). Even if all channels are used and modelled accurately information may be used to specify emissivity parameters, skin temperature or other surface characteristics rather than to provide information on atmospheric temperature. The impact is inevitably that less atmospheric temperature information can be extracted from the observations.

Despite these difficulties English (1999) found that temperature sounding using microwave radiometry is less sensitive to surface emissivity errors than sounding using infra-red instruments. Furthermore atlases of emissivity have been derived from SSM/I (Prigent et al. 1997) and these can be applied to improving temperature analyses from AMSU (English et al. 2000).

7. SUMMARY

Temperature profile information can be obtained from microwave radiance measurements between 50 and 57 GHz. The information has coarse vertical resolution but can provide useful information for temperature analysis, especially in the upper troposphere and stratosphere. The radiances can be sensitive to changes in variables other than atmospheric temperature, notably surface effects and cloud for the channels sensing at frequencies below 54 GHz. If an accurate temperature analysis is to be achieved these effects need to be taken into account either by appropriate quality control or modelling. If the radiances are to be used to generate a temperature analysis for numerical weather prediction this is best achieved by assimilating the radiances directly rather than to assimilate retrieved profiles.

NOAA15 AVHRR IR & AMSU cloud index composite imagery   LOCAL DATA

Image generated: 06/26/2000 11:46:07 UTC   Most recent overpass time: 1118 (Slot time: 1100)

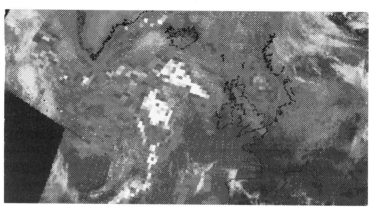

Figure 6: AMSU cloud liquid water detection. White areas indicate large amounts of liquid water detected.

8. REFERENCES

Derber J.C. and Wan-Shu Wu, 1998: The use of TOVS cloud-cleared radiances in the NCEP SSI analysis system. Monthly Weather Review, 126, 2287-2299.

English S.J., R.J. Renshaw, P.C. Dibben and J.R. Eyre 1997: The AAPP module for identifying precipitation, ice cloud, liquid water and surface type on the AMSU-A grid. Tech. Proc. 9th International TOVS Study Conference; Igls, Austria; 20-26 February 1997; Ed.: J.R. Eyre; Published by ECMWF, Reading, UK; 119-130.

English S.J. and T.J. Hewison, 1998: A fast generic millimetre-wave emissivity model, Proceedings of SPIE, 3503, 288-300.

English S.J., 1999: Estimation of temperature and humidity profile information from microwave radiances over different surface types., J. Appl. Meteorol., 38, 1526-1541.

English S.J., D.C. Jones, P.C. Dibben, R.J. Renshaw and J.R. Eyre, 1999b: The impact of cloud and precipitation on ATOVS soundings. Tech. Proc. 10th International TOVS Study Conference; Boulder, Colorado; 26 January-2 February 1999, 184-190.

English S.J., C. Poulsen and A.J. Smith, 2000: Forward modelling for liquid water clouds and land surface emissivity. Workshop on the use of ATOVS data for NWP assimilation, Published by ECWMF, Reading, RG2 9AX, 91-96.

Eyre, J.R. and A.C. Lorenc, 1989: Direct use of satellite sounding radiances in numerical weather prediction. Meteorol. Mag. 118, 3-16.

Eyre J.R., G.A. Kelly, A.P. McNally, E. Andersson and A. Persson, 1993: Assimilation of TOVS radiances through one dimensional variational analysis., Q. J. Royal Meteorol. Soc., 119(514), 1427-1463.

Eyre J.R., 1997: Variational assimilation of remotely-sensed observations of the atmosphere. J. Meteorol. Soc. Japan, 75, 331-338.

Gadd A.J., B.R. Barwell, S.J. Cox and R.J. Renshaw, 1995: Global processing of satellite sounding radiances in a numerical weather prediction system., Q. J. Royal Meteorol. Soc. 121, 615-630.

Grody N.C. and R.R. Ferraro, 1992: A comparison of passive microwave rainfall retrieval methods. Proc. 6th Conference on Meteorology and Oceanography, Atlanta, Georgia. Published by the American Meteorological Society, 60-65.

Grody N.C., F. Weng and R.R. Ferraro, 1999: Application of AMSU for obtaining water vapour, cloud liquid water, precipitation, snow cover and sea ice concentration. Tech. Proc. 10th International TOVS Study Conference; Boulder, Colorado; 26 January-2 February 1999, 230-240.

Kelly G., E. Andersson, A. Hollingsworth, P. Lonnberg, J. Pailleux, Z. Zhang, 1991: Quality control of operational physical retrievals of satellite sounding data. Mon. Wea. Rev. 119, 1866-1880.

Kelly G., 1999: Influence of observations on the operational ECMWF system., Tech. Proc. 9th International TOVS Study Conference; Igls, Austria; 20-26 February 1997; Ed.: J.R. Eyre; Published by ECMWF, Reading, UK; 239-244.

Petty G.W. and K.B. Katsaros, 1994: The response of the SSM/I to the marine environment. Part II. A parameterization of the effect of the sea surface slope distribution on emission an reflection., J. Atmos. and Oceanic Tech., 11(3), 617-628

Prigent C., E. Mathews and R. Rossow, 1997: Microwave land surface emissivities estimated from SSM/I observations. J. Geophys. Res., 102, 21867-21890.

Renshaw R.J., 2000: ATOVS asssimilation at UKMO. Workshop on the use of ATOVS data for NWP assimilation, Published by ECMWF, Reading, RG2 9AX, 45-50.

Rodgers C.D., 1976: Retrieval of atmospheric temperature and composition from remote measurements of thermal radiation., Revs. of Geophys. and Space Phys., 14, 609-624.

Rodgers C.D., 1990: Characterization and error analysis of profiles retrieved from remote sounding measurements. J. Geophys. Res., 95, 5587-5595.

Rosenkranz P.W., 1975: Shape of the 5mm oxygen band in the atmosphere, IEEE Trans. Ant. Prop., 23, 498-506.

Rosenkranz P.W., 1988: Interference coefficients for overlapping oxygen lines in air. J. Quant. Spectrosc. Radiat. Transfer, 39, 287-297.

Saunders R.W, 1993: A note on the Advanced Microwave Sounding Unit., Bulletin of the American Meteorol. Soc., 74(11), 2211-2212.

Saunders R.W., M. Matricardi, P. Brunel, 1999: An improved fast radiative transfer model for assimilation of satellite radiance observations. Q.J. R. Meteorol. Soc., 125, 1407-1426.

Thoss A., R. Bennartz and A. Dybbroe, 2000: Development of precipitation and cloud products from ATOVS/AVHRR data at SMHI. Workshop on the use of ATOVS data for NWP assimilation, Published by ECWMF, Reading, RG2 9AX, 167-174.

Microwave Limb Sounding

STEFAN BÜHLER
University of Bremen, Germany

Key words: Limb sounding, microwave, millimetre, sub-millimetre, MAS, MLS, MASTER, SOPRANO, SMILES, Odin.

Abstract: Microwave limb sounding is a well established technique for the observation of stratospheric trace gas species and is also receiving growing attention as a technique for the observation of tropospheric trace gas species. The article introduces the most important past and future instruments of this type, as well as some basics of the technique. A comprehensive list of references will make it possible for the interested reader to obtain more detailed information, both on microwave limb sounding in general and on the instruments discussed.

1. PAST AND FUTURE LIMB SOUNDING INSTRUMENTS

Microwave limb sounding is a well established technique for the observation of stratospheric trace gas species such as ozone and chlorine monoxide. It has been exploited by instruments such as the Microwave Limb Sounder (MLS) and the Millimeterwave Atmospheric Sounder (MAS). In recent years the receiver technology has advanced considerably, which makes it possible to envisage instruments of a far more ambitious character. The new generation includes MASTER (Millimeter Wave Acquisitions for Stratosphere/Troposphere Exchange Research), SOPRANO (Sub-Millimeter Observation of Processes in the Atmosphere noteworthy for Ozone), and JEM/SMILES (Superconducting Sub-Millimeter Wave Limb Emission bounder on the Japanese Experimental Module of the International Space

Station). It also includes the EOS/MLS (Microwave limb Sounder as part of the Earth Observing System) and Odin, an instrument shared between astronomers and upper atmosphere researchers. (The name Odin is not an acronym, but the name of a character from the nordish mythology.)

The first generation of limb sounders operated exclusively in the millimeter spectral range, but the second generation will operate in the millimeter and the sub-millimeter spectral range. The bandwidths are also now much larger, up to a few GHz, due to technological advances. This means that the new generation of limb sounders can measure all the way down into the upper troposphere. Finally, where the instruments of the first generation had system noise temperatures of a few thousand Kelvin, the JEM/SMILES instrument of the new generation will have a system noise temperature of only a few hundred Kelvin.

Advanced sensors require similarly advanced data evaluation or retrieval techniques, in order to cope with the higher data rate and better measurement precision. The optimal estimation technique is particularly well suited for simulation studies because it is generally applicable and because it facilitates a formal error analysis. The framework of that technique is defined in *Rodgers* [1976, 1990, 2000].

The extension of the measurement altitude range into the upper troposphere requires an analysis of the effect of continuum emissions. Furthermore, the larger bandwidth and — in the case of JEM/SMILES — lower noise temperature require an analysis of the impact of uncertainties in spectroscopic parameters. Both of these issues are addressed in *Bühler* [1999]. In the context of upper tropospheric measurements, an analysis of the impact of cirrus clouds is also crucial. A first assessment has been attempted in *Bühler et al.* [1999], but further work on this subject is necessary.

2. THE TECHNIQUE

2.1 Basics

The microwave limb sounding technique was originally developed in order to satisfy the need of stratospheric trace gas measurements with a good altitude resolution. A good introduction to this classical application of limb sounding can be found in *Waters* [1993]. Nowadays, the investigation of the tropopause region is a new application that has its own advantages and limitations.

Microwave limb sounding

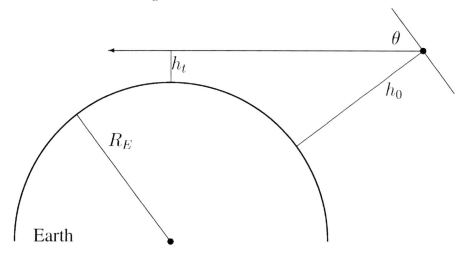

Figure 1: A schematic drawing of the limb sounding geometry. R_E is the Earth's radius. The receiver is in orbit at an altitude h_0 and looks tangentially through the atmosphere under an angle θ. The beam comes closest to the surface at the tangent point at an altitude h_t.

Points that are important in this context will receive more attention here than the more general aspects of limb sounding.

A limb sounder is a receiver that looks tangentially through the atmosphere, as indicated schematically in Figure 1. It is equipped with a movable antenna that makes it possible to adjust the viewing direction so that different tangent altitudes can be selected. The received radiation is analyzed by a spectrometer. In the case of a microwave instrument the spectral resolution can in principle be designed to be arbitrarily fine, but for practical reasons there is a tradeoff between spectral resolution and total bandwidth. The planned MASTER instrument will have a bandwidth of approximately 8 GHz per observation band (there will be 5 observation bands altogether), and a spectral resolution of the order of a few MHz.

The limb sounding technique can be used with or without an external source. In the case of a microwave instrument no external source is needed because the thermal emission from the atmosphere itself is sufficiently strong. The intensity of trace gas emission lines contains information about the trace gas concentration. The width of the lines contains information about the pressure, which can be used to determine precisely the instruments viewing angle.

Last but not least, the overall intensity of the signal, as well as the relative strengths of the different lines, contain information about the temperature. However, this information is mainly of interest as auxiliary data. For high accuracy temperature measurements there are better methods, either passive infrared sensing, or active microwave sensing using GPS signals.

2.2 Limits

The upper limit of the usable altitude range of a limb sounding measurement is determined by two factors, decreasing signal strength and limited spectral resolution. The signal decreases with altitude because the line intensity is proportional to the trace gas concentration which in turn is proportional to the total pressure (assuming that the volume mixing ratio stays constant). This means that the integral under an emission line decreases exponentially with altitude. The rate of the decrease of the emission for any given frequency depends on whether that frequency is located on a line wing or on a line center. In the line wing the decrease is exponential with altitude, in the line center it is only linear, because the line also gets narrower as the pressure decreases.

This narrowing of the lines with decreasing pressure is the reason why spectral resolution is the other limiting factor at high altitudes. Although it was said before that the resolution of a microwave instrument can in principle be arbitrarily fine there still is a practical limit because higher resolution comes at the price of increased noise. This can be seen from the *radiometer formula* that describes the noise characteristic of a microwave receiver:

$$\Delta T = \frac{T_S}{\sqrt{W t}} \quad (1)$$

where ΔT is the noise per channel, W is the channel width, t is the integration time, and T_S is the *system noise temperature*, a quantity that characterizes the 'noisiness' of a given receiver.

At what altitude the signal gets too weak for detection depends very strongly on the molecular species because line strengths — as well as mixing ratios — vary by many orders of magnitude. Pressure broadening, on the other hand, is relatively uniform; pressure broadening parameters only range from roughly 1 to 4 MHz/hPa. Therefore, a table of typical line widths makes it possible to estimate the spectral resolution that is necessary for measurements at different altitudes. The planned SOPRANO instrument, for example, has a nominal resolution of 3 MHz which would mean, according to Table 1, that the altitude range where the line shape can be spectrally resolved is up to 50 or 60 km. It is possible to study altitudes for which the line shape is not spectrally resolved, but altitude information in that case comes from the scanning geometry alone, so that the altitude resolution of the measurement will be degraded.

The *lower* limit of the measurement is more interesting in the context of upper tropospheric limb sounding, because one wants to extend the measurement down as far as possible and hence is continuously operating near the limit. As in the case of the upper limit, the lower limit is determined

Table 1: Typical pressure broadened line widths (half maximum) at different altitudes. At high altitudes (>50 km) the total width of the line is not any more determined by pressure broadening alone, but also by Doppler broadening which depends on frequency and temperature, but not on pressure.

altitude [km]	pressure [hPa]	γ(HWHM) [MHz]
0	1013.0	2569.0
10	281.0	842.0
20	59.5	188.0
30	13.2	39.7
40	3.33	9.33
50	0.951	2.53

by two factors, one physical and one instrumental. The physical limit is given by the high tropospheric opacity which eventually makes the limb path opaque. The main source of the high opacity is water vapor, hence, the absorption coefficient rises sharply in the vicinity of the tropopause. Continua — both water vapor and dry air — are extremely important in this context, since they are responsible for a large part of the total absorption coefficient. Continua are a lot more complicated than line absorption, both theoretically and experimentally. In the data analysis they are usually handled by adding fit parameters for an empirical absorption offset (for example a second order polynomial in frequency).

In addition to the physically limiting factor of rising opacity there is also the instrumentally limiting factor of finite receiver bandwidth. Table 1, which has already been useful in estimating the required *resolution* at high altitudes can also be used to estimate the required *bandwidth* at low altitudes. At an altitude of 5 km the typical full width at half maximum is 3 GHz. If one takes into account that pressure broadening of water vapor is stronger than that of most other species it follows that a bandwidth of up to 10 GHz may be necessary for tropospheric water vapor measurements. This requires not only a well designed receiver, but the band location has also to be carefully chosen because lines of other species in the vicinity also decrease the effective bandwidth that can be used for the water vapor measurement.

There is yet another factor that limits the low altitude measurement — not so much the altitude range but rather the temporal coverage. This factor is the effect of cirrus clouds on the retrieval. First model calculations indicate that weak clouds have no effect but strong clouds can affect the retrieval considerably [Bühler et al., 1999]. Luckily, the abundance of such 'strong clouds' is below 15 %, even at its maximum in the tropics.

2.3 Geometry, Antenna, and Resolution

It is now time for some geometrical considerations. Figure 1 defines the most important geometrical parameters. The receiver is in orbit at an altitude h_0 and looks tangentially through the atmosphere under an angle θ. The beam is lowest at the tangent point at an altitude h_t. One immediately sees that these parameters satisfy the equation

$$\cos\theta = \frac{R_E + h_t}{R_E + h_0} \qquad (2)$$

where R_E is the radius of the Earth. An important geometric quantity is the rate of change of h_t with θ, which is easily obtained by solving Equation 2 for h_t and taking the derivative:

$$\frac{dh_t}{d\theta} = -(R_E + h_0)\sin\theta \qquad (3)$$

Sometimes it is useful to have this differential as a function of h_t rather than θ. Solving Equation 2 for θ and taking the derivative leads to

$$\frac{d\theta}{dh_t} = \left[(2R_E + h_0 + h_t)(h_0 - h_t)\right]^{-1/2} \qquad (4)$$

The differential $dh_t/d\theta$ is for example needed in order to estimate the effect that the finite angular resolution of the antenna has on the measurement. Antennas are characterized by the *antenna response function* $A(\theta,\varphi)$, which states how strongly radiation from the direction (θ,φ) is attenuated with respect to the nominal looking direction $(\theta = 0, \varphi = 0)$. This function has many different names, depending on what units and which normalization are used. Common are, among others, *antenna pattern*, *field of view* (or FOV), or *antenna gain*. For the evaluation of limb sounder data the atmosphere is usually assumed to be homogeneous in the horizontal direction, so that the antenna response can be integrated in the φ direction and regarded as a function of θ only, where θ is defined as in Figure 1.

There are two integrated quantities that describe the most important antenna properties quite well, the full width at half maximum $\Delta\theta$ of the response function, also called *beam width*, and the *beam efficiency*, which is defined as the fraction of the signal collected from within a certain angular range —usually somewhat arbitrarily defined as 2.5 * $\Delta\theta$ —relative to that collected from all angles, for an isotropic radiation field. The beam width $\Delta\theta$ is a good measure for the angular resolution, a crucial parameter for limb sounder antennas, as will be discussed below. It is limited by the antenna diameter D_a, a rule of thumb is given in Equation 1.33 of *Janssen* [1993]:

$$\Delta\theta = 1.5\lambda / D_a \qquad (5)$$

where λ is the wavelength. The scaling factor depends on the antenna design, but for microwave antennas it is usually between 1.3 and 1.7. For limb sounding a narrow beam width is the most important antenna property, but it is also crucial that the beam efficiency is high because the troposphere and the ground emit thermal radiation that is an order of magnitude more intense than the radiation from the stratosphere.

Coming back to the question of resolution, what one is ultimately interested in is of course not the angular resolution of the antenna but the corresponding altitude resolution. This is readily obtained from

$$\Delta h_t = \frac{dh_t}{d\theta} \Delta\theta \qquad (6)$$

where Δh_t is the full width at half maximum of the antenna response in tangent altitude, and $dh_t/d\theta$ is the inverse of the differential defined in Equation 4. A summary of geometric and antenna parameters for the water vapor channels of MAS, MASTER, and SMILES is given in Table 2 (See Section 1 for a description of these instruments). To summarize, the vertical extent of the field of view at the tangent point is determined by the three factors platform altitude, antenna size, and frequency. It gets smaller with lower platform altitude, larger antenna size, and higher frequency.

The width of the field of view is important for the altitude resolution of the measurement, but the altitude resolution is generally narrower than the field of view because the pressure broadened line shapes also contain altitude information. The achievable altitude resolution is of the order of 1.5 km.

Radiative transfer calculations for limb sounders have to be done in two steps. First the transfer is calculated for single rays (also often called *pencil beams*) for a fixed angular grid. The spacing of this grid must be considerably smaller than $\Delta\theta$ in order to achieve the necessary accuracy. In the next step this set of radiances is convolved with the antenna response function, in order to simulated the effect of the antenna.

The discussion of the resolution in the vertical naturally raises the question of the resolution in the horizontal. Along the line of sight a good figure of merit is the extent of the layer that gives rise to half the total limb radiance for an optically thin ray path.

	h_0 [km]	$dh_t/d\theta$ [km/°]	D_a [m]	$\Delta\theta$ [°]	Δh_t [km]
MAS 183 GHz	300	34.0	1.0	0.1386	4.71
MASTER A200 GHz	800	57.1	2.2	0.0655	3.74
MASTER C 325 GHz				0.0437	2.50
SMILES 325 GHz	400	39.5	0.6	0.1304	5.15
SMILES 548 GHz				0.0774	3.06

Table 2: Geometric characteristics of different sensors: h_0 is the platform altitude, h_t is the tangent altitude, θ is the scan angle, D_a is the antenna diameter, $\Delta\theta$ is the full width at half maximum of the antenna response in θ, and Δh_t is the full width at half maximum of the field of view in tangent altitude. The values for $dh_t/d\theta$ and Δh_t depend on h_t. Here, a value of $h_t = 10$ km was assumed. For SMILES $\Delta\theta$ was calculated from D_a, using $\Delta\theta = 1.5\lambda / D_a$ (see text).

This has been calculated by *Waters* [1993] to be 1.6 km in the vertical and 286 km in the horizontal, under the assumption that the absorption coefficient is proportional to pressure. One could conclude that the horizontal resolution along the line of sight is also of the order of 300 km. However, if the instrument scans in the orbit plane, the same air parcels are sampled by different scans, which gives additional horizontal information. Then the horizontal resolution can be better than 100 km.

Perpendicular to the line of sight the resolution for a single scan is of the same order of magnitude as the field of view in the vertical direction, at least for a circular antenna. However, a more appropriate parameter is the sampling interval perpendicular to the line of sight, which is given by the orbit spacing if the instrument looks ahead or aft, or by the combination of scan duration and orbital velocity if the instrument looks sideways. The timing of the scan can be chosen such that the sampling interval is roughly equal to the resolution along the line of sight.

3. INSTRUMENT RESOURCES

MAS – The Millimeterwave Atmospheric Sounder

An instrument description can be found in *Croskey et al.* [1992]. Trace gas measurements are described in *Hartmann et al.* [1996]. That special issue of JGR also contains a number of other MAS publications. Temperature Retrieval is described in *Wehr et al.* [1998] and *von Engeln et al.* [1998]. Upper tropospheric humidity retrieval is described in *Bühler* [1999]. Web reference: http://www.iup.physik.uni-bremen.de/sat/projects/mas.

MLS — The Microwave Limb Sounder

MLS has yielded major scientific contributions in the areas of stratospheric ozone and chlorine chemistry research (high latitude ozone depletion), impact of volcano eruptions on the stratosphere (SO_2 loading), and tropical dynamics ('tape recorder' effect in the tropical lower stratosphere).

Barath et al. [1993] contains an instrument description. There are approximately 200 MLS publications from 1992 until now. Publication overview at: http://mls.jpl.nasa.gov/joe/um_pubs.html.

MASTER and SOPRANO

MASTER and SOPRANO are instrument concepts currently investigated by The European Space Agency ESA. Maybe MASTER will be part of a post-ENVISAT chemistry core mission. The name 'Millimeter Wave Acquisitions for Stratosphere / Troposphere Exchange Research' already says quite a lot about the instrument MASTER. However, MASTER will not only provide valuable data for stratosphere / troposphere exchange (STE) research, but also for research in the areas of radiative forcing, climate feedback, lower stratospheric chemistry, and upper tropospheric chemistry.

On the other hand, the name 'Sub-Millimeter Observation of Processes in the Atmosphere noteworthy for Ozone' indicates that the main purpose of SOPRANO is to collect data which is needed for stratospheric ozone chemistry research. For more details see *Bühler* [1999], which contains many more references. This book (and some other publications) can also be downloaded from: http://www.sat.uni-bremen.de/publications/.

JEM/SMILES

JEM/SMILES (Superconducting Sub-Millimeter Wave Limb Emission Sounder on the Japanese Experimental Module of the International Space Station) is the only currently planned satellite instrument that will have a superconducting (SIS) mixer. This makes the experiment highly challenging from a technological point of view. However, the SIS technology offers a lower instrument noise by a factor of 10, so that higher resolution O_3 measurements and more accurate ClO measurements will be possible. Details on the instrument can be found in *Masuko et al.* [1997]. Web reference: http://www.sat.uni-bremen.de/projects/smiles/.

EOS/MLS

This is the next generation of MLS. An overview is given in *Waters et al.* [1999]. Web reference: http://mls.jpl.nasa.gov/joe/eos_mls.html.

Odin

Odin, the first of the new generation limb sounders, was launched in February 2001. Details can be found in *Eriksson* [1999], and on the web page: http://www.ssc.se/ssd/ssat/odin.html.

4. REFERENCES

Barath, F. T., et al., The upper atmosphere research satellite microwave limb sounder instrument, *Journal of Geophysical Research*, 98, 10751–10762, 1993.

Bühler, S., *Microwave Limb Sounding of the Stratosphere and Upper Troposphere*, Shaker Verlag, Aachen, 1999, ISBcN 3-8265-4745-4.

Bühler, S., A. von Engeln, J. Urban, J. Wohlgemuth, and K. Künzi, The impact of cirrus clouds, in *Reburn et al.* [1999].

Croskey, C. L., et al., The Millimeter Wave Atmospheric Sounder (MAS): A Shuttle–based remote sensing experiment, *IEEE Trans. Microwave Theory Techn.*, 40, 1090–1100, 1992.

Eriksson, P., Microwave radiometric observations of the middle atmosphere: Simulations and inversions, Ph.D. thesis, Chalmers University of Technology, Göteborg, 1999.

Hartmann, G. K., et al., Measurements of O_3, H_2O, and ClO in the middle atmosphere using the Millimeter–wave Atmospheric Sounder (MAS), *Geophys. Res. Lett.*, 23, 2313–2316, 1996.

Janssen, M. A., An introduction to the passive microwave remote sensing of atmospheres, in *Atmospheric Remote Sensing by Microwave Radiometry*, edited by M. A. Janssen, chap. 1, John Wiley and Sons, Inc., 1993.

Masuko, H., S. Ochiai, Y. Irimajiri, J. Inatani, T. Noguchi, Y. Iida, N. Ikeda, and N. Tanioka, A superconducting sub-millimeter wave limb emission sounder (SMILES) on the Japanese experimental module (JEM) of the space station for observing trace gases in the middle atmosphere, in *Eighth International Symposium on Space Terahertz Technology*, Harward Science Center, 1997.

Reburn, W. J., et al., Study on upper troposphere / lower stratosphere sounding, final report, Tech. rep., ESTEC / Contract No 12053 / 97 / NL / CN, 1999.

Rodgers, C. D., Retrieval of atmospheric temperature and composition from remote measurements of thermal radiation, *Reviews of Geophysics and Space Physics*, 14, 609–624, 1976.

Rodgers, C. D., Characterization and error analysis of profiles retrieved from remote sounding measurements, *Journal of Geophysical Research*, 95, 5587–5595, 1990.

Rodgers, C. D., Inverse methods for atmospheric sounding, *World Scientific*, 2000, ISBN 981-02-2740-X.

von Engeln, A., S. Bühler, J. Langen, T. Wehr, and K. Künzi, Retrieval of upper stratospheric and mesospheric temperature profiles from Millimeter–Wave Atmospheric Sounder data, *Journal of Geophysical Research*, 103, 31735–31748, 1998.

Waters, J. W., Microwave limb sounding, in *Atmospheric Remote Sensing by Microwave Radiometry*, edited by M. A. Janssen, chap. 8, John Wiley and Sons, Inc., 1993.

Waters, J. W., et al., The UARS and EOS microwave limb sounder experiments, *Journal of Atmospheric Science*, 56, 194–218, 1999.

Wehr, T., S. Bühler, A. von Engeln, K. Künzi, and J. Langen, Retrieval of stratospheric temperatures from space borne microwave limb sounding measurements, *Journal of Geophysical Research*, 103, 25997-26006, 1998.

Retrieval of Integrated Water Vapour and Cloud Liquid Water Contents

LAURENCE EYMARD

CETP/IPSL/CNRS - Université St. Quentin-Versailles, Vélizy, France

Key words: microwave radiometry, atmospheric water vapour, cloud liquid water

Abstract: Since the mid seventies, spaceborne microwave radiometers have been used to get information about the atmosphere above oceans. Retrieval of water vapour content was performed first with the ESMR (Electrically Scanning Microwave Radiometer) flown on the Nimbus-5 satellite, then operational products from the SMMR (Scanning Multichannel Microwave Radiometer) onboard SEASAT and Nimbus-7 were routinely delivered (Wilheit and Chang, 1980, Prabhakara et al, 1983). The bases for establishing the retrieval of water vapour and cloud liquid water contents were thus established at the end of the seventies and beginning of the eighties, and most of the improvements achieved to date concern the tools used to perform the retrieval and the performances of the sensor. The SSMI (Special Sensor Microwave Imager) series has been flown continuously since 1987 to provide several products over oceans. Thanks to the good stability and calibration of its seven channels (between 19 and 85 GHz), many studies were performed to improve and evaluate the quality of the retrieval methods for atmosphere and ocean surface. Other microwave radiometers have been used to retrieve water vapour and liquid water contents, both on satellite and on ground, the latter helping to assess the spaceborne retrievals, as well as to analyse in details atmospheric transmittance modelling at microwave frequencies. Research are now developing on the retrieval of the water vapour profile, by using sensors with high frequencies as SSMT-2 and AMSU

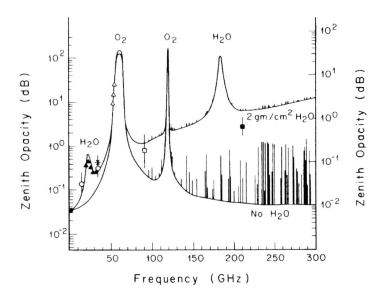

Figure 1: Atmospheric opacity for a US standard atmosphere. The lower line is with no water vapour, and the upper line is for a 20 kgm^{-2} WV content (assuming an exponential decrease with height). From Waters, 1976.

1. RADIATIVE TRANSFER MODELLING AND INVERSION METHODS

The atmospheric opacity spectrum (Figure 1) shows a first absorption line centred at 22.235 GHz, and a second one at 183.31 GHz.

Between these two lines, the water vapour continuum slowly increase absorption by the atmosphere with frequency. Two oxygen lines appear also in this frequency range, which are used for temperature profiling (see the section on atmospheric temperature analysis by S. English). The first water vapour line is too low to permit profiling, and its partial transparency is used to obtain the total columnar content. There is no absorption line for cloud droplets, but a continuous increase of scattering effects with frequency.

1.1 Radiative transfer modelling

The atmosphere is assumed horizontally stratified, and the radiative transfer equation in non-precipitating atmosphere for given frequencyν, polarisation p, and incidence andgle θ, may be very simply written as :

$$T_b(\nu,p,\theta) = e_s T_s \Gamma + T_a(1-\Gamma)(1+(1-e_s)\Gamma)+(1-e_s)\Gamma^2 T_C$$

with Ts the surface temperature, e_s the surface emissivity, $\Gamma=\exp(-\tau\sec\theta)$ the atmospheric integrated transmittance, Ta the equivalent air temperature (assuming an isothermal atmosphere) and Tc the cosmic background temperature.

The surface temperature and emissivity must be known for calculating the brightness temperature in the 1 – 100 GHz range, because of the weak opacity of the atmosphere. The ocean surface emissivity is small (of the order of 0.6) and depends on the surface roughness and temperature for each incidence angle, polarisation and frequency. A weak dependence on surface salinity exists, but is negligible above a few GHz. Many models have been developed to express the emissivity as a function of the wind just above the surface, allowing us to compute brightness temperatures rather accurately (see chapter 3). On the contrary, the land surface emissivity is close to unity, and varies with the soil dielectric properties, its content in water and the vegetation cover. Strong variations are often observed at small scales, making until now impossible to model the land emissivity with an accuracy sufficient for studies of the non-precipitating atmosphere. We will discuss later the feasibility of retrieving atmospheric contents over land.

The optical depth τ is the sum of the optical depths for oxygen, water vapour and clouds. The accuracy of transmittance modelling in clear atmosphere is now quite good (see chapter 1), and we will not discuss it any more. Most of studies on water vapour retrieval use the model developed by Liebe (1989, 1993). Concerning clouds, as long as the droplets diameter is small with respect to the wavelength, the Rayleigh scattering may be used, instead of the complete Mie scattering. In case of non-precipitating clouds, this assumption is verified in most cases in the 1-40 GHz range, allowing to strongly simplify the radiative transfer calculations and cloud description : the Rayleigh scattering is independent of the droplet size distribution, and results in an increase of the brightness temperature, similarly to absorption by water vapour and oxygen. This assumption is usually kept in the range 85-90 GHz, but is not as accurate as at lower frequency : Bobak and Ruf (2000) examined the sensitivity of brightness temperature computation through non-precipitating clouds with varying the cloud model (droplet size distribution and cloud structure) and the scattering calculations (Rayleigh approximation / Mie). They concluded that the Rayleigh approximations significantly deviates from the rigorous calculation for thick clouds with a total water content ≥ 1 kgm^{-2} at 85 GHz. Since such high contents are found generally in precipitating clouds, we may neglect these cases. However that light rain cannot be distinguished from clouds, and is implicitly included in the LWC retrieval.

1.2 Inversion methods

The general problem of inversion is discussed in chapter 1, and we will focus on aspects specific to retrieval of water vapour and cloud liquid water contents. With respect to other retrieval problems, as precipitation rate or surface characteristics, it is a weakly non-linear problem, and the assumption of a uniquely solution is verified. For upward looking radiometers, the accuracy of the retrieved water vapour content has been shown similar to the one of radiosondes, if not better. From space, the major difficulty is to eliminate the surface signal, which is much higher than the atmospheric one. A number of inversion methods have been proposed, using various combinations of frequencies and polarisations (horizontal / vertical).

- Statistical methods : The starting point is a data base containing both brightness temperatures (at the chosen frequencies and polarisations) and the corresponding water vapour and liquid water contents (hereafter noted as WV and LWC, respectively). The data base must be representative of all possible meteorological conditions either on a global or a regional scale. Such a method can easily be used for W, by collocating ship and small island radiosonde and /or ground based microwave radiometers with satellite overpasses. Due to the lack of any routine measurement of cloud water, it is quite impossible to build a globally representative data base for retrieving LWC. A possibility is to use operational meteorological model output fields, which include a diagnostic or prognostic cloud scheme. The inversion is then performed using a statistical method, with or without an a priori choice of channels or functions of the channels. As the WV retrieval is close to linear, a simple multilinear regression on brightness temperatures has often be used. The most wellknown purely statistical algorithms are those of Alishouse et al (1990a, b) for water vapour, which is a linear combination of 19, 22 and 37 GHz channels, with addition of a quadratic term in T22 to better account for the non-linearity, and for LWC, based on ground based microwave radiometer LWC estimates (linear function of brightness temperatures).

- Statistical inversion on a synthetic data base : First an atmospheric profile and surface data base (temperature $T(z)$ and humidity profiles $q(z)$, surface pressure ps, surface temperature Ts, 10m-wind Us) is selected, then a cloud model is used to simulate realistic cloud LWC on atmospheric profiles. Except for thick clouds, the retrieval does not significantly depend on the vertical distribution of droplet concentration and droplet size distribution, so a simple model is sufficient for SSMI-type sensors (Bobak and Ruf, 2000). Then brightness temperatures are computed using a radiative transfer model, and a statistical method (regression, neural network,...) is applied to establish the algorithm. The advantage of this method is that it is independent of the sensor, and thus a representative data base may be easily built. However, application to actual measurements generally requires some adjustments, due to radiative transfer model and radiometer calibration errors. Among this category, Wilheit et al (1980) established WV and CLW algorithms, using a multilinear regression. They showed that the nonlinearity of the transmittance (exponential form) can be reduced by using logarithmic functions of brightness temperatures. Karstens et al (1994) used similar functions of the brightness temperatures, as well as Gérard and Eymard (1998), using ECMWF predicted fields for building the learning data base. Other statistical algorithms (of linear form or with addition of polynomial correcting terms) were proposed by Bauer and Schlussel (1993), Weng and Grody (1994), among others, based on radiosonde profiles with a cloud model.

- Physical / physical-statistical methods : the estimate of the variable to be retrieved is adjusted using a radiative transfer model. The iterative process, starting from a first guess, is continued until the difference between simulated and observed brightness temperatures are lower than a given threshold. Radiative transfer models are generally simplified. When the errors (model, observations) are taken into account, these methods are called optimal estimation algorithms. A number of proposed retrievals use such a physical method : for example, Greenwald et al (1993, 1995) used two channels (19 and 22 for WV, 19 and 37 for CLW). They simplified the radiative transfer model by taking climatologic surface emissivity values. This method is close to the one developed by Wentz (1983, 1986) : first the wind speed is estimated with an empirical model, then the atmospheric transmittance is calculated at 22 and 37 GHz, in order to separate the contributions of water vapour and LWC (see also Liu et al, 1992). Phalippou (1996) used the assimilation scheme of a meteorological model to retrieve the WV and CLW : the first guess is the model analysis at the collocated point, and a 1-D variational assimilation procedure is applied to all SSMI channels. Other methods are based on a selection of channels or combinations of channels, in order to reduce the model uncertainties and optimise the retrieval : Petty and Katsaros (1992) established a model of the depolarisation effect of droplets on the sea surface signal in window channels (37 / 85 GHz). Other methods require additional information, on the atmospheric and surface temperature, cloud top (Liu and Curry (1993) among others).

1.3 Feasibility of WV and LWC retrieval over land

Very few studies have been performed : Jones and Vonder Haar (1990) showed the sensitivity of the SSMI 85.5 GHz channel to LWC, and Greenwald et al (1997) developed this idea, proposing two retrieval methods, using either one channel or a normalised polarisation ratio (difference of the two 85 GHz channels divided by the corresponding surface emissivity differences). These retrievals are based on a radiative transfer model, in which the surface emissivity is estimated in clear with additional information on the atmosphere water vapour and temperature. Prigent and Rossow (1999) analysed the theoretical sensitivity of all SSMI channels to surface temperature, WV and LWC. They stressed the importance of the emissivity range in the retrieval accuracy. A simultaneous retrieval of the surface temperature, WV and LWC is achieved by application of a variational inversion, using a NCEP analysis as first guess. They concluded that the WV could be retrieved with an acceptable accuracy ($3 - 4$ kgm^{-2}) only on clear sky and low emissivity areas (deserts), whereas the LWC can be obtained with an estimated accuracy of about 0.1 kgm^{-2}, high clouds being better detected than low clouds. As in the Greenwald et al (1997) study, the qualitative comparison with visible/IR cloud classifications is generally satisfactory, except for small or patchy clouds.

1.4 Error analysis

Errors in the retrieved products comes from several sources :

1.1.1 **The inversion method** : statistical inversion operates by minimising the global statistical deviation between the retrieved and the original value in the training data base. A multilinear regression thus tends to optimise the error in the centre of the distribution, whereas some bias occur at the two edges; neural networks, non-parametric methods improve the global fit, by better adjusting the non-linear relations, but the scatter in the training data distribution centre is slightly higher. Errors on iterative / optimal estimation methods mainly depend on the knowledge on the radiative model and observation errors, including any possible correlation.

1.1.2 **The radiative transfer model** : although the atmospheric transmittance modelling has greatly improved in the last 20 years, some uncertainties remain on the water vapour continuum. Cloud modelling (droplet concentration and size distribution) is much more uncertain, again, due to the lack of representative measurements in all cloud types. But the major uncertainty comes from the surface emissivity. Ocean surface emissivity modelling require a good knowledge on the sea state (related to the wind and swell, still an research subject), a model for scattering by small waves, and finally a model for foam cover and foam emissivity. As noted earlier, land surface modelling is still under study at all microwave frequencies.

1.1.3 **The training data base** : when using measured atmospheric profiles, the major problem is the representativity of all climate types (in particular, there is almost never any meteorological observation in the South-East Pacific Ocean, and very few in the major part of the southern hemisphere oceans). Their second problem is the quality of the measurements in very dry and very wet conditions : insufficient sensitivity of sondes, and saturation problems in clouds have been evidenced. Connell and Miller (1994) reported an average 1 kgm^{-2} error in the WV due to radiosonde relative humidity error. Moreover, radiosonde profiles transmitted to the meteorological network do not contain informations on the surface and on clouds (except the estimate of base and top from humidity). These limitations on representativity and quality propagate to meteorological models, but the quality control procedure and the assimilation scheme tend to reduce the observation errors. However, the quality of the model itself (parameterizations, grid resolution,...) and erroneous observations may cause systematic biases and noise. In particular the spin-up of the hydrological cycle during the first simulation hours may induce some biases in the WV and LWC. Concerning clouds, there is no direct method to assess the quality of estimates (either cloud simulations on observed profiles or cloud fraction and water contents predicted by meteorological models). Moreover, some cloud schemes, diagnosed from the saturation levels, may cause unrealistic correlations between the WV and LWC.

1.1.4 **The sensor calibration** : the absolute calibration of spaceborne microwave radiometer should be about 2 – 3 K (after ground calibration measurements). However, conditions in space are different, as for example the thermal emission from the close environment, and the microwave components may drift with time (and at launch). For these reasons, cross - comparisons between similar instruments, as well as comparisons between radiative transfer simulations and observations lead to differences up to 8 K.

As long as they are biases or linear trends, these difference should be taken into account before applying the inversion method, to prevent the retrieved products from being biased. In case of pure statistical methods, there is no effect of calibration errors, but the algorithm is not "transportable" to another sensor, even its characteristics are similar.

1.5 Validation / intercomparison of retrieved products.

Intercomparison of retrieval performances is often controversial, as noted by Petty (1999), because the claimed RMS error or correlation coefficient depend not only on the algorithm itself, but on a number of variables : two validation studies may yield different RMS errors, due to the differences in the validation data set (different criteria for data selection, for example). Moreover, the obtained error is the addition of the actual algorithm error (including possible sensor calibration error) and the validation data set error, which includes the noise due to imperfect collocation and local measurement error.

However, for the reasons described in 1.4.3, assessment of WV is much easier than for LWC. One of the first comparisons with radiosonde was made by Prabhakara et al (1983) with the SMMR retrieved WV. Jackson and Stephens (1995) and Deblonde and Wagneur (1997) reported results of intercomparison of several SSMI retrieved WV over the global oceans. They concluded that all algorithms overestimate the low humidity contents, whereas thay all underestimate the high values. The maximal discrepancy between them was found in the medium range $(20 - 30 \text{ kgm}^{-2})$. Other studies examined the behaviour of some SSMI algorithms in particular climatic locations or meteorological situations as did for example Eymard (1999) in the winter north Atlantic, Prigent et al (1997) in the vicinity of Azores islands during ASTEX, Alishouse et al (1990a) using radiosonde data from a selection of small islands. Most of published WV algorithms and retrievals have RMS differences with external data in the range $2 - 3 \text{ kgm}^{-2}$. Comparison with meteorological model WV fields shows very similar performances, as shows for example Prigent at al (1997) (Figure 2). Intercomparison with ground-based microwave radiometers have also be reported. As most of the published WV retrievals are consistent, there is no best choice, and simple algorithms as Alishouse et al (1990a) or Petty (1994) can be used for most of the applications.

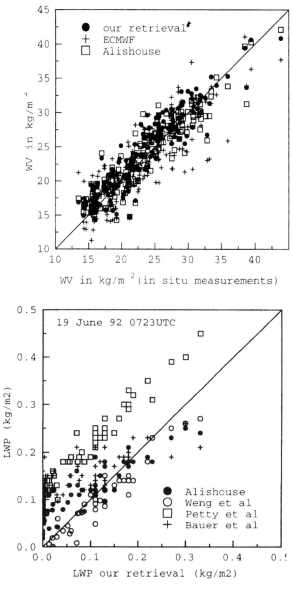

Figure 2 : Comparison of SSMI WV and LWC retrievals with in situ measurements and ECMWF WV fields during the ASTEX campaign (from Prigent et al , 1997). Left upper panel : WV comparison of the authors' variational retrieval, Alishouse (1990a) algorithm and ECMWF data with radiosondes. Left bottom panel: statistical comparison of several WV algorithms with in situ measurements. Right upper panel : Comparison of several LWC algorithms with the authors's retrieval. Bottom right panel : statistical performances of the 5 retrievals compared with the ISCCP estimate.

TABLE 1. Statistical differences between the integrated water vapor derived from radiosonde measurements and the estimation by various SSM/I water vapor algorithms (the mean, standard deviation, and roor-mean-square of the differences are presented). Also presented are statistical difference between in situ measurements and ECMWF estimates. The comparison involved 173 radiosonde measurements.

	Our retrieval	ECM-WF	Alishouse et al. (1990)	Schluessel et al. (1990)	Petty et al. (1990)	Bauer et al. (1993)
Mean	-0.7	0.3	-0.2	1.6	1.2	-1.3
Std.dev.	2.1	3.3	2.3	2.3	2.4	2.4
rms	2.3	3.3	2.3	2.8	2.7	2.7

LWC intercomparison and validation is still more risky, since independent measurements are sparse (aircraft flights in clouds during a few experiments) and are not representative of the cloud variability and LWC range. One of the major problems with such comparisons is the heterogeneity of the atmosphere at the satellite pixel scale : a high number of aircraft profiles should be necessary to account for the statistical properties of clouds, but the air flow stationarity is not long enough (1 – 2 hours in general) for permitting a consistent comparison with a satellite overpass.

TABLE 2. Mean and standard deviation of the LWP estimates (kg M-2) for the ISCCP procedure and for the SSM/I retrievals. Both clear and cloudy pixels are included in the calculation; 2680 pixels are involved. The numbers on the left represent the results when including the negative LWP values; the values in parentheses on the right indicate the some quantities but when setting all the negative SSM/LWP retrievals to zero (when negative values are present).

	ISCCP	Our retrieval	Weng et al. (1994)	Alishouse et al. (1990)	Petty et al. (1990)	Bauer et al (1993)
Mean	0.03	0.02	0.01(0.03)	0.03(0.04)	0.10	0.12
Std.dev.	0.04	0.05	0.06(0.05)	0.07(0.06)	0.08	0.05

LWC intercomparison and validation is still more risky, since independent measurements are sparse (aircraft flights in clouds during a few experiments) and are not representative of the cloud variability and LWC range. One of the major problems with such comparisons is the heterogeneity of the atmosphere at the satellite pixel scale : a high number of aircraft profiles should be necessary to account for the statistical properties of clouds, but the air flow stationarity is not long enough (1 – 2 hours in general) for permitting a consistent comparison with a satellite overpass.

Comparison studies are therefore often made also with visible/IR retrieval methods and ground-based radiometer estimates, but they are not any more free of space-time mismatches. Weng and Grody (1994), Eymard (2000) compared SSMI LWC with ground based radiometer estimates (respectively on small islands and ship). Gérard and Eymard (1998) compared SSMI LWC with aircraft profiles and with a Meteosat cloud classification to assess the consistency of the retrieval in clear air and stratiform cloud conditions. Prigent at al (1997) compared several SSMI retrievals with aircraft measurements and ISCCP estimates (Figure 2). A more complete comparison was performed in the framework of the CLOREVAL study (Cloud Retrieval Validation experiment, Offiler et al (1998)). Nineteen SSMI retrieval algorithms were compared with aircraft measurements in stratiform clouds taken during the ACE-2 experiment and a dedicated Cloreval experiment in the North Sea. Careful screening of the data was made to reduce the mismatches and scaling errors. Among the 19 algorithms, seven selected for their good performances with respect to aircraft measurements. Despite they are based on different principles (they include the Alishouse et al (1990b) statistical algorithm, some statistical retrievals on synthetic data bases (Gérard and Eymard, 1998, Karstens et al, 1994, Krasnopolsky, 1997, Jung et al, 1998, the two latter using a neural network instead of a multilinear regression for the two former), and a physical retrieval (Greenwald et al, 1995). However, the low range of measured LWC ($0 - 0.2$ kgm^{-2} mainly) and the moderate values of corresponding WV explain the good performance of the Alishouse algorithm, despite it was found to have an anomalous sensitivity to WV (Petty and Katsaros, 1990, Penney, 1994). Penney (1994) made a different and complementary intercomparison, by applying several LWC algorithms to simulations of SSMI brightness temperatures made on radiosonde profiles, and discussing their sensitivity to other parameters (cloud temperature, WV, surface wind) in various climatic situations.

From the above various comparisons, it results that the physical retrieval method proposed by Greenwald et al (1993, 1995) works satisfactorily in the range $0 - 0.5$ kgm^{-2}. Among the statistical algorithms, those using a combination of $\log(T0-Tbi)$, T0 being a constant (280 or 290 K), and Tbi the brightness temperature of the ith channel, work better than linear combinations of brightness temperatures. There is no evidence that there is a optimal choice of channels for LWC retrieval, since the published algorithms use 2 to 7 SSMI channels! From the various comparisons cited previously, the accuracy of LWC (given by an RMS error with respect to a reference data set) ranges between 0.04 and 0.1 kgm^{-2}. A minimum error of about 0.05 kgm^{-2} should therefore be considered for any application.

Above 0.5 kgm^{-2}, most of the algorithms underestimate the LWC. An improved behaviour can be obtained by using neural network or non-parametric methods instead of a linear regression. But the significance of LWC above 0.5 kgm^{-2} is questionable, since the probability of rain increases. However, high values (up to 1 kgm^{-2}) were obtained in non-precipitating clouds using surface-based radiometers by Eymard (2000) and Snider

(2000). The difference with space-borne measurements lies in the different observation scales and smoothing. Eymard (2000) showed that the internal variability (expressed as the LWC standard deviation) of a given cloud is proportional to its mean LWC value, whatever be the cloud sampling.

2. APPLICATIONS OF WV AND LWC RETRIEVALS

The rather good quality of WV retrieval has led to a large variety of applications. Some of them include also use of the retrieved LWC.

2.1 Assimilation in forecast models and validation of atmospheric models

Following first assimilation experiments (Filiberti et al, 1994, 1998), the SSMI derived WV was shown to significantly impact the model fields, not only humidity but also the atmospheric circulation. In the ECMWF model, the WV is first retrieved using the 1-D variational scheme (Phalippou, 1996) then it is incorporated in the 4-D variational assimilation scheme, as other observations. Variational assimilation increases significantly the sensitivity to humidity in the model, with respect to optimal interpolation : Filiberti et al (1994) showed that the humidity information brought by the WV is rapidly lost in optimal interpolation, because there is no backward effect on the other fields. On the contrary the global 3-D var or 4-D var schemes (Rabier et al, 1997) minimises, in a least square sense, the difference between the model and all observations available on a given period.

Retrieved products from microwave radiometers have also been used to evaluate the performances of meteorological model analyses and forecasts (Liu et al, 1992, Deblonde et al). Such comparison could evidence regional anomalies of the model hydrological cycle, as well as errors induced by the assimilation of other sensors (TOVS or GOES humidity data).

2.2 Low-level humidity and surface latent heat flux

Liu (1984) demonstrated the potential of using spaceborne microwave radiometers to estimate the surface latent heat flux over oceans, through application of a simple bulk formula (relationship between the turbulent flux and the low level wind and humidity and the humidity at the surface, function of the surface temperature). The major step in this method is to obtain the low atmosphere humidity. Liu established from radiosonde data that the low-level humidity can be related to the WV, at a monthly scale. The corresponding 5th order polynomial provides the specific humidity with an accuracy of 0.8 gkg^{-1} (Liu, 1986). The same idea was developed by several authors, at various time and space scales, and Liu et al (1991) showed statistically that the atmospheric boundary layer is strongly related to the total precipitable water at scales over 10 days, whereas Schlussel et al (1995) directly derived the low-level humidity from SSMI brightness temperatures. However, Esbensen et al (1993) performed an error analysis of Liu's method, and showed that the major contribution comes from the low level humidity, leading to biases in the evaporation flux both in the tropics and at high latitudes. Bourras and Eymard (2000) re-examined the estimate of surface heat fluxes using SSMI data (and additional information for the surface temperature), showing the feasibility of a direct statistical relationship (neural network) between the surface latent heat flux and SSMI brightness temperatures. The resulting RMS error is about 40 Wm^{-2} at the global and regional scale (comparison with ship flux measurements) without any time averaging.

2.3 Altimeter tropospheric correction

Altimetry missions (Topex/Poseidon, ERS/1-2, future Jason-1 and ENVISAT) have the primary objective to precisely monitor the sea surface height, with respect to the earth geoid. The method employed is to measure the path duration of the radar pulse emitted by the altimeter and reflected by the surface. Unfortunately, a number of phenomena affect the path delay and must be either modelled or independently measured. In particular, the tropospheric WV is routinely measured using a 2- or 3-channel microwave radiometer, pointing vertically as the altimeter. Such a companion instrument was shown necessary to correct for abrupt changes in WV that are not properly forecasted by meteorological models, despite the assimilation of SSMI WV. Keihm et al (1995) and Eymard et al (1996) proposed retrieval algorithms for the tropospheric path delay due to water vapour. They are conceptually identical to WV algorithms, to which they are proportional (by a factor of 6). LWC algorithms were also derived by these authors, in order to correct for the radar signal attenuation. The performances of the Topex and ERS vertically pointing radiometers are similar to each other, although the ERS/MWR is a two-channel radiometer (23.8 and 36.5 GHz) : the effect of the surface is provided by including a function of the altimeter backscatter coefficient (or the derived surface wind) in the statistical retrieval. However, Gérard and Eymard (1998) found that the LWC algorithm for ERS is more scattered than the one developed for SSMI, because of the smaller number of channels, and the uncertainty on the altimeter wind.

3. CONCLUSIONS AND PERSPECTIVES

The spaceborne microwave radiometers have proved their capability for accurately retrieving the columnar water vapour, and are the most reliable technique to obtain the cloud liquid water content over the oceans. Many retrieval methods have been proposed in the last 20 years, from simple statistical regressions to variational inversion. For WV, no method has been shown significantly better than the others, all yielding an accuracy of 2 – 3 kgm^{-2}. A reason of this limitation is the accuracy of the "ground truth" generally consisting of a set of radiosonde profiles from ship or small islands, another one is the scatter generated by time/space mismatches between in situ measurements and satellite footprints. For LWC, additional problems come from the scarcity of reliable LWC measurements, and from the high heterogeneity of cloud shape and structure. However, recent intercomparisons showed a satisfactory consistency of various retrieval methods in case of nearly uniform clouds.

Progresses in this domain will mainly come from new sensors or combination of sensors :

- Microwave multichannel imagers with an enhanced horizontal resolution, as TMI (TRMM Microwave Imager), due to its low altitude, and more significantly AMSR on board ADEOS and the near similar radiometer on board the EOS-AQUA platform. In addition, these instrument will be accompanied by high quality visible/IR radiometers, offering new perspectives in combined retrievals. This increased resolution will also perhaps improve the accuracy of retrievals over land.
- Microwave humidity profilers, as the SSMT-2 and AMSU-B sensors : radiative transfer modelling and humidity profile retrieval is been established for the 183.31 GHz water vapour line and associated window channels (Kuo et al, 1994, Wilheit and Hutchinson, 1997, Engelen and Stephens, 1999 among others) and this retrieval is operational in Meteorological Centres. Some studies have focused on the sensitivity of high frequency brightness temperatures to cloud liquid and ice contents (e.g. Wang et al, 1997, Muller et al, 1994). Miao (1998) has shown the feasibility of retrieving the total precipitable water over polar ice with SSMT-2 data. The good horizontal resolution of the AMSU-B will probably lead to new developments, in particular over land and ice, in complement to the new imagers. The simultaneous presence of AMSU-B and AMSR on board EOS-AQUA will be an opportunity to develop

combinations of the two sensors, benefiting from their large range of frequencies and good horizontal resolution.

4. REFERENCES:

Alishouse, J.C., S.A. Snyder, J. Vongsathorn and R.R. Ferraro, 1990a: Determination of oceanic total column water vapour from the SSM/I. IEEE Transactions on Geoscience and Remote Sensing, 28, 5, 811–816.

Alishouse, J.C., J.B. Snider, E.R. Westwater, C.T. Swift, C.S. Ruf, S.A. Snyder, J. Vongsathorn and R.R. Ferraro, 1990b: Determination of cloud liquid water-content using the SSM/I. IEEE Transactions on Geoscience and Remote Sensing, 28, 5, 817–822.

Bobak, J.P., and C.S. Ruf, 2000 : Improvements and complications involved with adding an 85 GHz channel to cloud liquid water radiometers. IEEE Trans. Geos. Remote Sensing, 38, 214-225.

Bauer, P., and P. Schluessel, 1993 : Rainfall, total water, ice water, and water vapor over sea from polarized microwave simulations and SSM/I data. J Geophys. Res., 98, 20737-20759.

Bourras, D. et L. Eymard : Comparison of three methods for retrieving the surface latent heat flux over oceans from SSMI. In "Microwave Radiometry and Remote Sensing of the Earth's Surface and Atmosphere", P. Pampaloni and S. Paloscia eds, VSP Int. Science Publishers,47-56, 2000

Connell, B.H., and D.R. Miller, 1995 : An interpretation of radiosonde errors in the atmospheric boundary layer. J. Applied Meteor., 34, 1070-1081.

Deblonde, G., and N. Wagneur, 1996 : Evaluation of of global numerical weather prediction analyses and forecasts using DMSP SSM/I retrievals. 1. Satellite retrieval algorithm intercomparison study. J. Geophys. Res., 102, D2, 1822-1850.

Deblonde, G., W. Yu, and L. Garand, 1997 : Evaluation of global numerical weather prediction analyses and forecasts using DMSP SSM/I retrievals. 2. Analyses/forecasts intercomparison with SSM/I retrievals. J. Geophys. Res., 102, D2, 1851-1866.

Esbensen, S.K., D.B. Chelton, D. Vickers and J. Sun, 1993 : An analysis of errors in Special Sensor Microwave Imager evaporation estimates over the global oceans. J.Geophys. Res. 98, C4, 7081 - 7101.

Engelen, R.J., and G.L. Stephens, 1999 : Characterization of water-vapour retrievals from TOVS/HIRS and SSM/T-2 measurements. Quaterly J. Meteor. Soc., 125, 331-351.

Eymard, L., C. Klapisz, and R. Bernard, Comparison between Nimbus-7 SMMR and ECMWF model analyses : The problem of the surface latent heat flux, J. Atmos. Oceanic Technol., 6 (6), 966-991, 1989

Eymard, L., R. Bernard, and J.Y. Lojou, Validation of microwave radiometer geophysical parameters using meteorological model analyses, Int. J. Remote Sensing , 14, 1945-1963, 1993

Eymard, L. : Analysis of cloud liquid water content characteristics from SSMI and a shipborne radiometer. In "Microwave Radiometry and Remote Sensing of the Earth's Surface and Atmosphere", P. Pampaloni and S. Paloscia eds, VSP Int. Science Publishers, 235-245, 2000

Filiberti, M.A., L. Eymard, and B. Urban, Assimilation of satellite precipitable water in a meteorological forecast model, Mon. Wea. Rev., 122, 3, 486 - 506, 1994

Filiberti, M.A., F. Rabier, J.N. Thépaut, L. Eymard, P. Courtier, Four-dimensional variational SSM/I water vapour data assimilation in the ECMWF model IFS, Quaterly J. Royal Meteor. Soc., 124, 1743-1770, 1998

Gérard, É. and L. Eymard, 1998: Remote Sensing of integrated cloud liquid water: development of algorithms and quality control. Radio Science, 33, 433–447.

Gérard, E., and R.W. Saunders, 1999, 4Dvar assimilation os SSM/I total column water vapour in the ECMWF model. Quaterly J. Royal Meteor. Soc.,

Greenwald, T.J., C.L. Combs, A.S. Jones, D.L. randel and T.H. Vonder Haar, 1997 : Further developments in estimating cloud liquid water over land using microwave and infrared satellite measurements. J. Applied Meteor, 36, 389-405

Greenwald, T.J., G.L. Stephens, T.H. Vonder Haar and D.L. Jackson, 1993: A physical retrieval of cloud liquid water over the global oceans using Special Sensor Microwave/Imager (SSM/I) observations. Journal of Geophysical Research, 98, 18471–18488.

Greenwald, T.J., G.L. Stephens, S.A. Christopher and T.H. Vonder Haar, 1995: Observations of the global characteristics and regional radiative effects of marine cloud liquid water. Journal of Climate, 8, 2928–2946.

Greenwald, T.J., C.L. Combs, A.S. Jones, D.L. Randel, and T.H. Vonder Haar, 1997: Further developments in estimating cloud liquid water over land using microwave and infrared satellite measurements. J. Applied Meteor., 36, 389-405.

Jackson, D.L., and G.L. Stephens, 1995 :A study of SSM/I derived columnar water vapor over the global ocean. J. Climate, 8, 2025-2038.

Jung, T., E. Ruprecht and F. Wagner, 1998: Determination of cloud liquid water path over the oceans from SSM/I data using neural networks. Journal of Applied Meteorology, 37, 832–844.

Karstens, U., C. Simmer and E. Ruprecht, 1994: Remote sensing of cloud liquid water. Meteorology and Atmospheric Physics, 54, 157–171.

Keihm, S., M. Janssen, and C. Ruf, 1995 : Topex/Poseidon microwave radiometer (TMR) : III. Wet troposphere range correction algorithm and pre-launch error budget. IEEE Trans. Geosci. Remote Sensing, 33, 147-161.

Krasnopolsky, V.M., 1997: A neural network-based forward model for direct assimilation of SSM/I brightness temperatures. NOAA-NCEP, Environmental Modelling Center, Ocean Modelling Branch, Technical Note No.140, 33pp.

Kuo, C.C., D.H. Staelin, and P.W. Rosenkranz, 1994 : Statistical iterative scheme for estimating atmospheric relative humidity profiles. IEEE Trans. Geos. Remote Sensing, 32, 254-260.

Ledvina, D.V., and J. Pfaendtner, 1995 : Inclusion of SSM/I total precipitable water estimates into the GOES data assimilation system. Mon. Weather Rev., 123, 3003-3015.

Liebe, H.J., 1989 : MPM - an atmospheric millimeter-wave propagation model. Intern. J. Infrared Millimeter Waves, 10, 631-650.

Liebe, H.J., G.A. Hufford and M.G. Cotton, 1993: Propagation modelling of moist air and suspended water/ice at frequencies below 100GHz. In Proceedings AGARD 52nd Specialists' Meeting of the Electromagnetic Wave Propagation Panel, Palma De Mallorca, Spain, 17–21 May 1993, 3-1-3-10.

Liu, G. and J.A. Curry, 1993: Determination of characteristic features of cloud liquid water from satellite microwave measurements. Journal of Geophysical Research, 98, 5069–5092.

Liu, W.T., 1984 : Estimation of latent heat flux with SEASAT-SMMR: a case study in North Atlantic. "Large scale Oceanographis Experiments and satellites", compilation by C. Gautier and M. Fieux. (Reidel), 205 - 221.

Liu, W.T., 1986: Statistical relation between monthly mean precipitable water and surface level humidity over global oceans. Mon. Weather Rev., 114, 1591-1602..

Liu, W.T., W. Tang and P.P. Niiler, 1991 : Humidity profiles over the ocean. J. Climate, 4, 1023-1034.

Liu, W.T., W. Tang and F.J. Wentz, 1992 : precipitable water and surface humidity over global oceans from SSM/I and ECMWF. J. Geophys. Research, 97, C2, 2251-2264.

Muller, B.M., H.E. Fuelberg and X Xiang, 1994 : Simulations of effects of water vapor, cloud liquid water, and ice on AMSU moisture channel brightness temperatures. J. Applied Meteor., 33, 1133-1154.

Offiler, D., 1998: CLOREVAL – SSM/I liquid water path validation – data. In Proceedings of AMS 9th Conference on Satellite Meteorology and Oceanography, Paris, 25–29 May 1998. Eumetsat EUM-P-22, 2, 646–649.

Penney, G., 1994: A sensitivity study of SSM/I statistical and physical liquid water path algorithms using a radiative transfer model. Technical note of the Meteorological Office no 110, 26 pp.

Petty, G.W. and K.B. Katsaros, 1990: New geophysical algorithms for the Special Sensor Microwave/Imager. Proceedings of the 5th Conference on Satellite Meteorology and Oceanography, London, American Meteorological Society, 247–251.

Petty, G.W. and K.B. Katsaros, 1992: The response of the SSMI to the marine environment. Part I : An analytic model for the atmospheric component of observed brightness temperatures. J. Atmos. Ocean. Tech., 9, 746-761.

Petty, G.W., 1994, Physical retrievals over ocean rain rate from multichannel microwave imagery. Part II : Algorithm implementation. Meteor. Atmos. Physics, 54, 101-121.

Petty, G.W., 1999, Review of retrieval and analysis methods for passive microwave imagers, In Microwave Radiometry for atmospheric research and monitoring, report of the COST 712 action, ESA/ESTEC Eds. No WPP-139, 3-29.

Phalippou, L., 1996: Variational retrieval of humidity profile, wind speed and cloud liquid-water path with the SSM/I: Potential for numerical weather prediction. Quarterly Journal of the Royal Meteorological Society, 122(530), 327–355.

Prabhakara, C., H.D. Chang and A.T.C. Chang, 1982 : Remote sensing of precipitable water over the oceans from Nimbus-7 microwave measurements. J. Appl. Meteor., 21, 2023-2037.

Prigent, C., L. Phalippou, and S. English, 1997, Variational inversion of the SSM/I observations during the ASTEX campaign. J. Applied Meteor, 36, 493-508

Prigent, C., and W.R. Rossow, 1999 : Retrieval of surface and atmospheric parameters over land from SSM/I : Potential and limitations. Quaterly J. Royal Meteor. Soc., 125, 2379-2400.

Rabier, F. et al, 1997 : Recent experimentation on 4D-var and first results from a simplified Kalman filter. ECMWF technical memorandum 240.

Ruf, C.S., S.J. Keihm, B. Subramanya, and M.A. Janssen, 1994 : TOPEX/POSEIDON microwave radiometer performance and in-flight calibration. J. Geophys. Res., 99, 24915-24926.

Schluessel, P. L. Schanz, and G. Englisch, 1995 : Retrieval of latent heat flux and longwave irradiance at the sea surface from SSM/I and AVHRR measurements. Advances in Space Research, 16, 107-116.

Snider, J.B., 2000, Long-term observations of cloud liquid, water vapor, and cloud base temperature in the north Atlantic ocean. J. Atmos. Ocean. Technol., 17, 928-939.

Wang, J.R., J. Zhan and P. Racette, 1997 : Storm-associated microwave radiometric signatures in the frequency range of 90 – 220 GHz. J. Atmos. Ocean. Technol., 14, 13-31.

Weng, F. and N.C. Grody, 1994: Retrieval of cloud liquid water using the special sensor microwave imager (SSM/I). Journal of Geophysical Research, 99, 25,535–25,551.

Weng, F., N.C. Grody, R. Ferraro, A. Basist and D. Forsyth, 1997: Cloud liquid water climatology from the Special Sensor Microwave/Imager. Journal of Climate, 10, 1086–1098.

Wentz, F.J., 1993 : A model function for ocean microwave brightness temperatures. J. Geophys. Res., 88, 1892-1908.

Wilheit T. T. and A. T. C. Chang, 1980: An algorithm for retrieval of Ocean Surface and Atmospheric Parameters from the Observations of the Scanning Multichannel Microwave Radiometer. Radio Sci., 15, 3, 525-544.

Wilheit T. T. and K.D. Hutchison, 1997 : Water vapour profile retrievals from SSM/T-2 data constrained by infrared-based cloud parameters. Int. J. Remote Sensing, 18, 3263-3277.

Laurence Eymard was born in 1954. She graduated from Ecole Normale Superieure and Université Paris6 in 1978. She joined the Centre de Recherche en Physique de l'Environnement (CRPE) where she became involved in research activities in the field of boundary layer meteorology, until 1985, then satellite microwave radiometry over oceans, with a particular interest to air - sea interactions. She received her doctorate degree (thèse d'Etat) in physics of the atmosphere in 1985. Since 1991, she has become the project scientist for the ERS-1 microwave radiometer, and she developed the retrieval algorithms and performed the in-flight calibration and validation of the two ERS microwave radiometers. In the same time she has become involved in experimental studies of the ocean-atmosphere interactions (SOFIA programme), and she participated in experiments in the North Atlantic (SOFIA/ASTEX, SEMAPHORE, CATCH/FASTEX) and Mediterranean sea (FETCH).She has a particular interest for the improvement of microwave instrument use over oceans for the retrieval of surface and atmospheric parameters (coupled inversion of measurements, surface model improvement), and relations with turbulent surface fluxes, as well as use of microwave radiometry in meteorologic models (assimilation, validation).

Precipitation Retrieval From Spaceborne Microwave Radiometers and Combined Sensors

FRANK S. MARZANO, ALBERTO MUGNAI AND F. JOSEPH TURK
Centre of Excellence CETEMPS - University of L'Aquila, L'Aquila, Italy
Institute of Atmospheric and Climate Sciences, CNR, Rome, Italy
Naval Research Laboratory, Monterey, California, USA

Keywords Precipitation, microwave radiometry, infrared radiometry, radar, spaceborne sensors, retrieval algorithms.

Abstract Applications of space-based microwave radiometry to precipitation retrieval is addressed, focusing on model-based inversion techniques and optimal combination with other satellite sensors. Statistical integration of microwave radiometers with space-borne radars and thermal infrared radiometers is sketched. Examples of measurements and products are shown to emphasize the physical background of combined techniques. The concept of the newly approved NASA/NASDA Global Precipitation Mission (GPM), considered to be the Tropical Rainfall Measuring Mission follow-on international space program on precipitation remote sensing, is finally illustrated together with concluding remarks on current status of precipitation retrieval.

1. INTRODUCTION

Precipitation is a key meteorological parameter, knowledge of which is required in a number of research and application disciplines directly related to the global water and energy cycle and its underlying rainfall processes (Houze, 1981). This includes climate diagnostics and modeling, numerical weather prediction, flood forecasting, hydrology, oceanography, agro-meteorology and water management. Since precipitation is a crucial element affecting human life on Earth and constitutes the foremost exchange process within the hydrological cycle, the need for accurate global measurements of rainfall amounts and distributions has become more and more important (Simpson et al., 1988; Tao et al., 1993). This is particularly true in an era of climatic uncertainty where predictions of freshwater resources and other major water-related phenomena have become paramount. Furthermore, the benefits of including (i.e., assimilating) rainfall information in global climate and numerical weather prediction (NWP) models has recently been demonstrated, and these successes have greatly increased user requirements for operational global precipitation datasets, since they are recognized to have prognostic value as well as more conventional value as aids to diagnostic analysis and model verification (e.g., Marecal and Mahfouf, 2002).

Notably, due to the great variability of precipitation in space and time, measurements with the necessary detail and accuracy to make these predictions are difficult to obtain and as a result, our current knowledge of global rainfall is unsatisfactory. Ground-based measurements – either by rain gauges or meteorological radars – provide only sparse data over land (especially, in mountainous regions) and very little data over the oceans. Thus, it is generally recognized that satellites are the only viable means for providing global and continuous precipitation data sets. To this end, passive microwave (MW) sensors are particularly suitable to this purpose (e.g., Mugnai et al., 1990; Smith et al., 1998).

New aspects and applications of airborne and spaceborne microwave radiometry have been also addressed in the last years. One major topic is related to its combination with space-borne radars (e.g., Olson et al., 1996; Marzano et al., 1999; Viltard et al., 2000; Bauer, 2001) or thermal-infrared radiometers (e.g., Levizzani et al., 1996; Vicente et al., 1998; Turk et al., 1998a; Marzano et al., 2001). For the most part, the activity has concerned convective precipitation, however, retrieval of stratiform precipitation has recently received a careful consideration both from a forward modeling and from an inversion point of view (e.g., Bauer et al., 2000; Marzano and Bauer, 2001).

In the following sections, we will briefly describe some aspects of these new features of passive microwave rain retrieval, referring to proper bibliography for further details. A final section will be briefly devoted to the concept of the Global Precipitation Mission (GPM), that has been recently approved by NASA and NASDA and it is now in its formulation stage as the Tropical Rainfall Measuring Mission (TRMM) follow-on international space program on global precipitation remote sensing from space. Some further comments and considerations will close this contribution.

2. SPACEBORNE MICROWAVE RADIOMETRY

Spaceborne and airborne microwave radiometry has proved to be a candidate tool for providing fairly accurate estimates of precipitation. Several inversion methods have been applied to extract rainfall rate from multi-frequency radiometric measurements (Wilheit et al., 1994; Smith et al., 1998).

Most retrieval algorithms use linear, non-linear or iterative inversion techniques based on an empirical or a physical approach (e.g., Mugnai et al., 1993; Bauer and Schluessel, 1993; Smith et al, 1994a; Kummerow and Giglio, 1994; Evans et al., 1995; Skofronick-Jackson and Gasiewskii, 1995; Marzano et al., 1994; Kummerow et al., 1996; Pierdicca et al., 1996; Turk et al., 1998b). The differences between the two approaches stems from the fact that they can be trained either by co-located ground-based and radiometric measurements or by one-dimensional (1D) or three-dimensional (3D) cloud models coupled with microwave radiative transfer schemes (e.g., Smith et al., 1994b; Ferraro and Marks, 1995; Conner and Petty, 1998). The various inversion approaches are aimed at providing, as a final product, the surface rainrate and/or the integrated parameters or the vertical profiles of cloud and precipitation hydrometeor contents. A preliminary discrimination of surface coverage and identification of rainy areas, possibly separated between stratiform and convective precipitation, is also of primary importance for applying the MW-based retrieval algorithms (Bauer and Grody, 1995; Hong et al., 1999; Marzano et al., 2001).

Nevertheless, several issues still remain open problems, especially in terms of absolute accuracy of rainfall estimates over ocean and, mainly, over land (Ferraro and Marks, 1995; d'Auria et al., 1998; Conner and Petty, 1998). With respect to empirical approaches based exclusively on spaceborne and ground measurements, physical inversion techniques can take advantage of the various degrees of freedom offered by a forward radiative-transfer model (Smith et al., 1994b). Their appealing feature is related to the possibility to investigate new spaceborne passive instruments and advanced retrieval techniques by acting in a self-consistent simulation environment. But, as a drawback, model-based approaches always need to be tackled with the representativeness of the models and the simulated data sets they produce, namely the cloud-radiation data set used to train the inversion algorithm (Mugnai et al., 1993; Marzano et al., 1994; Panegrossi et al., 1998; Bauer, 2001; Mugnai et al., 2001; Tassa et al., 2002). The requirement for a better representativeness of the cloud-radiative data set in the specific area of interest demands for a tuning to local geographical and climatological conditions (Kummerow et al., 2001).

As for the passive MW measurements from Low Earth Orbit (LEO) satellites, a rapid evolution is expected within the next few years. Currently, two Special Sensor Microwave/Imager (SSM/I) radiometers are operational on the sun-synchronous polar orbiting satellite series of the U.S. Defense Meteorological Satellite Program (DMSP). The DMSP program consists of continuously flying two satellites at fixed equatorial crossing times (5:30 and 8:30 a.m.). In this two-satellite system, the SSM/I is presently being replaced by the Special Sensor Microwave Imager/Sounder (SSMIS) which includes temperature and humidity sounding channels (at about 50-60 GHz and 183 GHz, respectively) in addition to the four standard SSM/I frequencies, corresponding to three atmospheric windows (at about 19, 37, and 85 GHz) and one water vapor absorption line at about 22 GHz. In addition, two Advanced Microwave Scanning Radiometers (AMSRs) -- providing brightness temperature (TB) measurements at two additional window frequencies (6.9 and 10.7 GHz) and with significantly higher horizontal resolutions than SSM/I and SSMIS -- are about to be flown on other two sun-synchronous polar orbiting satellites: NASDA's Advanced Earth Observation Satellite – II (ADEOS-II) and NASA's Earth Observing System (EOS) Aqua mission, having 10:30 and 13:30 equatorial crossing times, respectively.. These four advanced MW radiometers will provide continued frequent coverage (at 4-5 hour intervals at mid-latitudes) for the next few years, within a data policy framework granting near-real-time access to the data.

Table 1: Comparison of past, current and future spaceborne MW radiometers.

Frequency (GHz)	SMMR (1978) IFOV (km)	polarization	SSM/I (1987) IFOV (km)	polarization	TMI (1997) IFOV (km)	polarization	SSMIS (2001) IFOV (km)	polarization	AMSR (2002) IFOV (km)	polarization
6.9	120	H, V							42	H, V
10.7	74	H, V			47	H, V			27	H, V
~ 19	44	H, V	56	H, V	26	H, V	56	H, V	15	H, V
~ 23	37	H, V	49	V	22	V	49	V	13	H, V
~ 37	21		30	H, V	13	H, V	30	H, V	7.9	H, V
~ 50 (2 channels)							39	H	5.5	V
~ 57 (6 channels)							39	H		
~ 60 (5 channels)							78	H		
~ 90			13	H, V	5.5	H, V	13	H, V	3.2	H, V
150							39	H		
~ 183 (3 channels)							39	H		

Finally, a special mention must be made to the MW measurements provided by the Tropical Rainfall Measuring Mission (TRMM), that was launched in 1997 by NASA and NASDA at a 350-km altitude, with a 35-degree inclination non-sun-synchronous orbit in order to conduct systematic measurements of tropical rainfall required for major studies in weather and climate research (Simpson et al., 1988; Kummerow et al., 1998, 2000). This rain observatory (which is expected to remain in operations into the 2005-06 period – due to the recent boost to a 402 km altitude) carries the TRMM Microwave Imager (TMI) radiometer (derived from SSM/I with the addition of one window frequency at 10.7 GHz) and the first space-borne rain radar, the Ku-band (13.8 GHz) Precipitation Radar (PR) – thus providing the most accurate precipitation measurements up to date. In order to appreciate the potential benefit of the advanced MW radiometers that will become available, Table 1 compares the main characteristics of SSMIS and AMSR with those of their precursors [SMMR is the Scanning Multichannel Microwave Radiometer, that was launched on Nimbus-7 and SeaSat in the 70s.].

3. COMBINED PASSIVE AND ACTIVE MICROWAVE RAIN RETRIEVAL

Many works have dealt with the comparison of active and passive microwave observations of precipitation, even though relatively few attempts have been made in order to combine radar and radiometer measurements within the same precipitation retrieval scheme (Weinman et al., 1990; Olson et al., 1996; Haddad et al., 1996; Meneghini et al., 1997; Marzano et al., 1999; Marzano and Bauer, 2001). Recently, however, this topic has been gaining an ever-increasing attention in the remote sensing community due the launch of the TRMM mission carrying aboard a combined MW radar/radiometer payload.).

The goal of a combined technique is to develop methods for integrating active and passive microwave data to better estimate the precipitation structure and the surface rainrate. On one hand, radiometer measurements can be used to properly account for the contamination of radar echoes due to fairly strong hydrometeor attenuation at frequencies above 10 GHz. On the other hand, the radar range resolution can help the retrieval algorithms to better define the near-surface precipitation vertical profile.

TRMM preparatory projects were an incredible source to design and test combined retrieval techniques. Within these field campaigns, it is worth mentioning the Tropical Ocean-Global Atmosphere - Coupled Ocean-Atmosphere Response Experiment (TOGA-COARE), that was carried out over the tropical south Pacific Ocean during November-December, 1992 and January-February, 1993 (Yuter et al., 1995; McGaughey et al., 1996). During TOGA-COARE, the Advanced Microwave Precipitation Radiometer (AMPR), having four channels between 10 GHz and 85 GHz, was flown aboard the ER-2 NASA aircraft together with the Millimeter-Wave Imaging

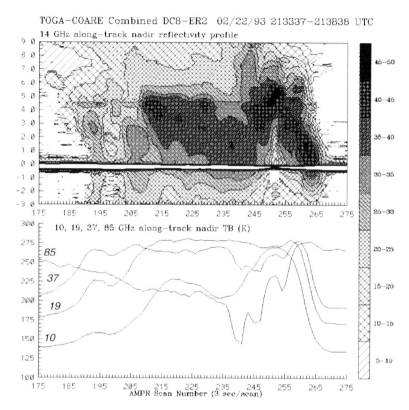

Figure 1. TOGA COARE measurements at nadir on February 22, 1993 in terms of reflectivity cross-sections (top) and corresponding upwelling brightness temperatures (bottom). AMPR T_B's at 85, 37, 19, and 10 GHz are correspondingly ordered from top to bottom at the initial scan 175 (adapted from Marzano et al., 1999).

Radiometer (MIR), having six channels between 89 GHz and 220 GHz. Moreover, several observations were coordinated with the 13.8 GHz JPL Airborne Rain Mapping Radar (ARMAR) aboard a DC-8 aircraft (Durden et al., 1994). The fact that the active and passive instrumentation was installed aboard two aircrafts flying at different altitudes limited the potential of the TOGA-COARE combined data set. However, the wide range of available channels from 10 GHz up to 220 GHz and the number of various observations of stratiform and convective rainfall made the campaign itself a unique opportunity to test new approaches to combined rainfall retrieval. In particular, the MIR measurements have given a quantitative appreciation of the potential of millimeter-wave frequencies above 90 GHz for rainfall retrieval -- an aspect that is gaining more and more attention in literature.

In our research, combined precipitation retrieval methods have been applied to nadir-looking radar and radiometric measurements relative to two

TOGA-COARE storm cases (Marzano et al., 1999). In Fig. 1, the observed case of February 22, 1993 is shown.

Noteworthy, within the convective core region (about from pixel 235 to pixel 265) the 37 GHz TB has a behavior more similar to the 85 GHz TB than to the 19 GHz TB, due to the overwhelming scattering effects at the higher frequencies. On the other hand, the maximum (around pixel 250) of 10 GHz TB is not coincident with the minimum (around pixel 240) of 85 GHz TB – an anomalous behavior, which is due to the tilted nature of the observed squall line and is explained by the fact that the 10 GHz TB is more sensitive to the lower-cloud rain layers, while the 85 GHz TB to the upper-cloud ice layers. Finally, note that the 10 GHz values show another relative peak around scan 220 due to a stratiform-like cloud structure.

Radar/radiometer combined precipitation retrieval techniques can be based on the Bayesian inversion approach (Olson et al., 1996; Marzano et al., 1999; Di Michele et al., 2002). The retrieval products consist of the estimated vertical profiles of rain, cloud water, graupel, and ice aggregate contents together with the surface rainrate. Specifically, the radiometer-based retrieval technique utilizing the upwelling TBs measured by both AMPR and MIR, can be further extended in order to manage the corresponding radar reflectivities both as independent measurements and as additional constraints to the radiometer estimation process. Noteworthy, a technique to approach the radar-swath broadening problem can be also developed within this framework (Di Michele et al, 2002).

An interesting feature of Bayesian retrieval techniques is that it is possible to insert *a priori* information in a rigorous way so as to preserve the formal description of the inverse problem (Marzano et al., 1999). In other words, the statistical inversion framework, based on the Bayesian criterion, allows one to perform a multi-parameter non-linear retrieval by using a cloud-radiation database as *a priori* constraint and to easily integrate measurements acquired from various remote sensors.

Simulated and experimental results have shown that a fairly good improvement is obtained when ARMAR data are added to the AMPR ones, especially for estimating surface rainrate, graupel profiles, and rain profiles. In addition, the analysis of both AMPR and ARMAR data and the application of a combined estimation algorithm have revealed significant features of the observed mesoscale convective systems -- in particular, the hydrometeor composition of the tilted structure of the squall line.

A test of combined retrieval algorithm can give an idea of the expected accuracies when using a radar together with a radiometer (Marzano and Bauer, 2001). Fig. 2 shows the root mean square (RMS) errors for the retrieval of water, graupel, snow, and ice profiles, obtained by performing a self-test at nadir over ocean. The solid curves refer to the use of radiometer data only (AMPR like), while the dash-dotted curves to radar data only

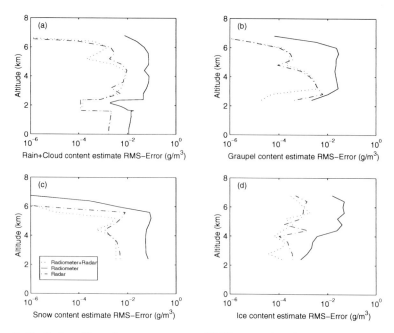

Figure 2. Vertical profiles of root mean square (RMS) errors for rain (a), graupel (b), snow (c), and ice (d) hydrometeor content estimates, obtained by performing a self-test at nadir over ocean. The solid curves refer to the use of radiometer data only (AMPR like: 10.6, 19.3, 37.0, and 85 GHz channels), the dash-dotted curves to radar data (ARMAR like: 36 range-gated reflectivities between 0 and 8 km plus total attenuation at 13.8 GHz), while the dashed curves to combined radiometer plus radar data (AMPR+ARMAR like). Only stratiform profiles have been considered (adapted from Marzano and Bauer, 2001).

(ARMAR like). The dashed curves apply to radiometer plus radar data(AMPR+ARMAR like).

As known, the contribution of Ku-band radar data helps the microwave radiometer in improving the accuracy of large and dense particle retrieval, such as raindrops and graupel. As seen from the figure, the improvement can be higher than 50%. The use of radar data only gives results slightly worse than the combined ones, especially for cloud liquid water above the freezing level FL and graupel content below 4 km where the radiometer information is more effective in the retrieval synergy.

4. COMBINED PASSIVE MICROWAVE AND INFRA-RED RAIN RETRIEVAL

Measurements from space-borne visible-infrared (VIS-IR) sensors are not directly related to precipitation microphysics since the upwelling radiation depends on near-cloud-top reflectance and emission (i.e., temperature).

Thus, VIS-IR precipitation estimates are only indirect and need to be calibrated (Adler et al., 1994; Levizzani et al., 1996). On the other hand, VIS-IR sensors provide high-resolution measurements even when aboard geo-stationary platforms – making it possible to follow storm development at the relevant time and space scales (Vicente et al, 1998; Turk et al., 1998a). Conversely, passive MW measurements are more directly related to precipitation microphysics since the upwelling TBs depend, in a frequency-dependent fashion, on the internal structure of precipitating clouds throughout the entire cloud and rain column (e.g., Smith et al, 1992). Thus, MW precipitation estimates can be physically based, (in principle) they do not need to be calibrated, and are more accurate than the VIS-IR estimates.

On the other hand, MW sensors require large antennas and generally have coarse ground resolutions even when flown on LEO satellites. As a consequence, current space-borne MW sensors do not provide the temporal sampling, spatial resolution and coverage needed to understand the different rainfall regimes around the world nor to provide sufficient input to prediction models (e.g., Kummerow et al., 1998). In order to take advantage of the respective strengths of the two types of measurement, combined MW/VIS-IR techniques may be developed and exploited, in which the MW measurements serve as a calibration source for VIS-IR measurements providing precipitation estimates at the wide spatial scale and rapid temporal update cycle of observations from geo-synchronous satellites (e.g., Turk et al., 1999; Marzano et al., 2001; Miller et al., 2001; Turk et al., 2002).

The statistical integration method is aimed at combining infrared radiometric data from geostationary platforms and microwave radiometric data from sun-synchronous platforms, by deriving probability-matched relationships between the histograms of the IR equivalent-blackbody temperatures and the SSM/I and TMI derived rainrates (Atlas et al., 1990; Crosson et al., 1996).

In this type of hybrid probability matching method (PMM), we rely upon the micro/millimeter-wave based sensors to supply the necessary physically-based instantaneous rainfall estimation observations, and the operational geostationary-based infrared imagers to supply the large-scale spatial and frequent update time history of these same cloud systems from an infrared (IR)-based perspective. Since the useful field of view of a geostationary satellite extends about 50 to 60 degrees about the satellite subpoint, the presence of different evolving rain systems in this domain must be accomodated. In this technique, we accomplish this by subdividing the Earth into 7.5-degree boxes between ± 60 degrees latitude. Previous passes of the appropriate geostationary IR data are time and space-aligned (10-minute maximum allowed time delta between the time of the IR and MW data) newest-to-oldest, using the geolocation data for each colocated data point to build up separate histograms of the IR temperatures and the associated rain rate for the appropriate box. To assure that only the most

Principles of Hybrid - Statistical Precipitation Analysis

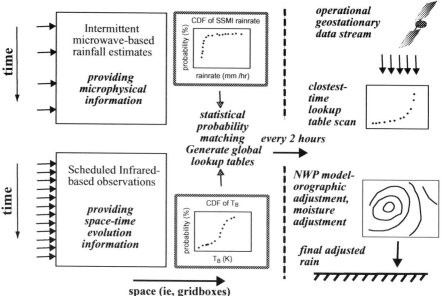

Figure 3. Flow chart of the procedure to combined spaceborne microwave radiometric and geostationary infrared data to estimate surface rainfall on a near real-time and global scale (adapted from Marzano et al., 2001).

recent rain history is captured, the rain rate and IR temperature histograms are accumulated until the percent coverage of a given box exceeds a coverage threshold, which we have set to 90 percent. Depending upon how recently a given geographical region was covered by an SSM/I, AMSU-B or TMI orbit, the overall data used in some of the histogram boxes may be only a few hours old, whereas other regions may require more look-back time to reach the coverage threshold. Currently, we set a maximum look-back time limit of 24 hours on the process. If a region has not reached the coverage threshold by this time, the histograms for this region are marked bad and the rain rates set to "no data" until the next update cycle. However, eventually an update cycle will capture newly-arrived microwave data that have arrived in this region. The 7.5-degree boxes are spaced every 2.5-degrees apart, so that they overlap geographically, assuring that the histograms from adjacent boxes transition smoothly from one to the next. Figure 3 depicts the process in a block-diagram flowchart. Comparisons with ground-based rain gauge data over Korea and Australia (Turk et al, 2002) has demonstrated correlations of about 0.65 for daily rainfall accumulations at scales of 0.25-degrees (about 25-km). At 3-hour time scales, the correlations fall off rapidly to near 0.3, since in this short time interval there are relatively few microwave-based data available to update the probabilistic histogram matching and the information content is mainly from the IR data.

Precipitation retrieval from spaceborne 117

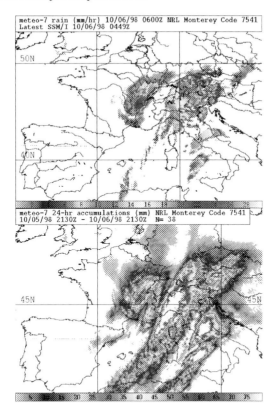

Figure 4. Maps of estimated surface rainrate (top) at 0600 UTC of October 6, 1998 and accumulated rain (bottom) from 2130 of October 5, 1998 till 2130 UTC of October 6, 1998 (adapted from Marzano et al., 2001).

Figure 4 shows an example of the application of the PMM technique to a precipitation event occurred on October 6, 1998 in Southern Europe. The SSM/I overpass was around 0600 UTC. Both the instantaneous at 0600 UTC and the 24-hour accumulated rainrate are shown. Note that the latter has been estimated by adding the rainrate estimated by PMM-calibrated Meteosat images available every half an hour, that is 48 rainrate maps (instead of 2 possible SSM/I-derived maps from a single DMSP platform).

Limits and perspectives of these combined MW-IR techniques can be, briefly, outlined (e.g., Marzano et al., 2001). The accuracy of the combined technique for estimating rainfall, in coincidence of an overpass of a sun-synchronous MW platform, is comparable to the one obtained from microwave radiometers. In particular, the expected accuracy of rainfall over land is generally worse than that over ocean. For what concerns the limitations of microwave radiometry in rain detection and estimation, sensors like SSM/I best capture convective-type rain over land where there is significant ice scattering. SSM/I ground-resolution of the order of tens of kilometers is also a problem when observing cumulonimbus clouds which can have an horizontal size less than 10 km, even when organized in multicells or supercells. Hence, these characteristics are reflected in the PMM performance.

When using the combined technique for "calibrating" the IR measurements by PMM, the expected accuracy generally suffers for impairments such as low-stratus detection and high-cloud screening. A comparison with raingage data, performed during February 1998 over western California, put in evidence an overestimation in the California Central Valley by a factor from 2 to 4 even when applying the high-cloud screening (Turk et al., 1999). The main reason of this disagreement is probably due to orographic effects which are not taken into account in the current implementation. However, the more important result is that, even though the quantitative estimate of rain may be not precise, the position of the intense storm cores is always well tracked. For flood alarms and fade mitigation techniques, this feature could be by itself a very valuable information. Actually, the validation of satellite-derived rainfall and attenuation is not an easy task to be accomplished since ground-based measurements are not always easily accessible and, apart from radar, raingage and earth-satellite stations can provide only point or line-of-sight data, respectively. Weather-radar estimates of rainfall would represent a more effective comparison, even though they are very often affected by calibration errors and attenuation effects when operating at C-band.

Proposed high-cloud screening techniques can be applied to Geostationary Operational Environmental Satellites (GOES) data, but not to the current Meteosat due to lack of a dual-channel in the thermal IR (Vicente et al., 1998). The Second Generation of Meteosat platforms (MSG), planned to be launched in year 2002, will prospect a new scenario to really improve this screening step, since it will have twelve different channels for cloud detection capabilities. The designed image cycle is of 15 minutes, the pixel sizes of 1.4 km and 4.8 km for VIS and IR channels, and with 12 channels for enhanced imaging and pseudo-sounding. In particular, the MSG will have 2 channels at 6.2 and 7.3 µm in the water-vapor absorption band, four channels at 3.9, 8.7, 10.8, and 12.0 µm in the IR windows for surface and cloud detection, and 2 channels at 9.7 and 13.4 µm for atmospheric pseudo-sounding in the CO_2 absorption bands. Most problems in rain cloud detection and characterization will be more effectively addressed by using MSG data, and the combined technique based on PMM can be easily extended to handle the new measurement set.

5. THE CONCEPT OF A GLOBAL PRECIPITATION MISSION

The NASA/NASDA Tropical Rainfall Measuring Mission (TRMM) was launched in November 1997 and has been since conducting systematic measurements of tropical rainfall, that are required for major studies in weather and climate research. TRMM is the first mission primarily dedicated to precipitation (and latent heat release) research and its observations of tropical cyclones have been extraordinarily useful not only for rainfall measurements, but also for quantifying the three-dimensional distribution of precipitation and (tentatively) of the associated latent heat release, and therefore for understanding the physical and dynamical structure of such systems (Simpson et al., 1988; Kummerow et al., 1998; 2000; 2001).

TRMM, however, has not solved all the problems associated with precipitation mapping and monitoring. Consistently with its design, TRMM rainfall uncertainties are dominated by the time-space sampling errors, as the inaccuracy due to insufficient temporal sampling cannot be overcome with a single LEO satellite. Also, the characteristics of the TRMM Precipitation Radar (non-Doppler, single frequency) make it difficult to retrieve directly the latent heat profile (which is a necessary input for accurate model predictions) and to determine the drop size distribution (which is a necessary parameter for accurate precipitation retrievals).

In addition, the TRMM low inclination (35 degrees) does not allow observations of the rainfall regimes at the mid and high latitudes. Also, due to the highly variable nature of precipitation, TRMM's crude sampling frequency effectively limits utilization of its data for short-term weather forecasting and numerical weather prediction applications. Finally, the data generated by TRMM may be well suited for intra-seasonal climate studies, but are hardly sufficient for long-term climate variability studies (TRMM is expected to be operational for about four more years), including the global warming trend, the El Nino life-cycle and frequency of recurrence, and the long-term climate cycles caused by deep ocean circulation. In essence, the major TRMM system drawbacks are summarized below.

- *Spatial coverage*: TRMM does not provide measurements outside of the tropics (35°N - 35°S).
- *Temporal sampling*: the sampling frequency at any given point by the TRMM radiometer is limited to approximately 1 sample every 15 hrs, while the TRMM radar is limited to approximately 1 sample every 50 hrs (depending upon the latitude of the sample).
- *Lifetime*: TRMM is expected to remain in operations until 2005-2006.

These are the main reasons why NASA and NASDA have been considering a TRMM Follow-on mission that will not only extend the TRMM precipitation climatology in the next decade, but will also address such issues. This mission, which has been called Global Precipitation Mission (GPM), will be launched in year 2007 and will be basically constituted by a primary platform and a satellite constellation (e.g., Mugnai and Testud, 2002).

The concept behind the GPM configuration is that the primary, or "core", satellite will provide detailed and accurate measurements of the three-dimensional distribution of precipitation and its associated latent heat release for most rainfall regimes around the globe, while the constellation of "drone" satellites will provide the temporal sampling that is required for most precipitation studies and applications.

The primary satellite should be an upgrade of the current TRMM satellite. High spatial resolution and rainfall accuracy will be reached by using a dual-frequency (14 and 35 GHz) imaging pulsed radar(possibly with Doppler capabilities for the higher frequency) in combination with a TMI-like radiometer. Since the primary satellite is basically designed to examine the instantaneous structure of precipitation and latent heating, it is acceptable to have variable gaps in the observations. Nevertheless, the core satellite will be in a non-sun-synchronous orbit to observe the diurnal variable rainfall structure, while its inclination will be around 65° in order to observe all the important rainfall regimes around the globe.

The constellation satellites, embarking basically small radiometers, are envisioned to be in near-polar, sun-synchronous orbits. Not only do these orbits afford the most effective way to sample the globe with a fixed repetition time, they also make it possible to immediately take advantage of already available radiometer data. Without contributions from existing platforms, 8 drone satellites will be needed to achieve 3 hourly sampling with the number growing to 24 satellites for 1 hour revisit times. The number of drone satellites, however, can be immediately reduced once the number of existing radiometers is factored in. Currently, there are 2 SSM/Is (or SSMIS) in orbit that can take the place of these drone satellites, but in the near future, there might be as many as four satellites (3 NPOESS and 1 AMSR) in orbit. International space agencies, as the European Space Agency (ESA), have been invited to contribute to the constellation satellites. Thus, a proposal for an European constellation satellite has been recently submitted to ESA by a large European and international team (Mugnai and Testud, 2002).

The core satellite should also help in properly calibrating rainfall estimates made by the drone satellites. The drone radiometers will provide the necessary sampling to reduce errors in time-averaged rainfall estimates to levels significantly smaller than the intrinsic errors in hydrologic and atmospheric models. The error due to temporal sampling should be reduced to approximately 10% for daily rainfall accumulations. Moreover, the drones must be very small, light, and low on energy consumption. The primary cost driver for these satellites should be the satellite bus and launch vehicle. By keeping the drone radiometers small and light, the cost of launching individual constellation satellites could be kept affordable. Note that GPM system is envisioned as an effort lasting at least 20 years.

From what mentioned above, it emerges that the objectives of GPM should overcome TRMM deficiencies by satisfying the following mission requirements:
1. global spatial coverage with a rainfall accuracy estimation better than that available by using current SSM/I radiometers aboard DMSP satellites;
2. temporal sampling less or equal to 3 hours in order to improve numerical weather prediction models, data assimilation models and hydrological models and, possibly, to help flash flood forecast;
3. life-time sufficiently long in order to monitor and understand potential long-term changes in the spatial structure of precipitation and related latent heating;
4. efficient down-link data transmission in order to provide flash flood forecast centers with near-real time products.

6. CONCLUDING REMARKS

As a conclusion, we wish to remark upon some peculiar aspects of current research on passive microwave remote sensing of precipitation from space.
1. Cloud-radiation model-based approaches have proved to be a viable way to tackle the rainfall retrieval problem from space. The use of mesoscale cloud-resolving model simulations can provide a physically consistent database of hydrometeor contents profiles. Available radiative transfer models can address most of the basic aspects of simulating the upwelling brightness temperatures from precipitating clouds. The effectiveness of this approach has been successfully proved by TRMM, since the operational precipitation algorithm has been founded on these considerations (Kummerow et al., 2001). In this context, Bayesian inversion methods have shown to be, both theoretically and numerically, a flexible tool for developing rainfall retrieval algorithms. However, some specific issues are still under investigation and it is worth mentioning them now:
    - The objective criteria to generate a statistically-significant training database for a global-scale application – i.e., (to a large extent) the problem of how to select and merge different model outputs whose underlined microphysics and dynamics can be very different from each other.
    - The modeling of melting ice particles – especially, but not only, for quantifying the impact of the melting layer on the upwelling TBs for stratiform precipitation. Such a problem is also related to the possibility of automatically separating convective rainfall from the

stratiform one by using satellite radiometric data with some a priori information.
- The modeling of hydrometeor non-sphericity – which would allow to quantify the effects of hydrometeor shape on the upwelling TBs and would enable an effective use of polarization diversity within precipitation retrieval algorithms.
- The accuracy of precipitation retrievals and their inherent uncertainties – especially over land, due to the usually larger and less known background emissivity (e.g., Petty, 1995).
- The optimal treatment of the different ground resolutions at the various frequencies of the measuring radiometer, in order to physically determine the ground resolution of the retrieved products – an issue which is related to the sensitivity of each frequency to different vertical portions of the cloud itself, as well as to the inclined view of a conically scanning radiometer.

In addition, extended and comprehensive validation strategies should be developed and planned by resorting to both ground-based and airborne data, as already done during the TRMM preparatory and on-going field campaigns. Finally, algorithm improvement should also be pursued by means of systematic inter-comparisons of the various model-based and empirical retrieval algorithms, as it was extensively done in the pre-TRMM era.

2. Microwave radiometry has demonstrated to have several advantages with respect to other techniques, such as infrared radiometry and microwave radar. Schematically speaking, with respect to infrared radiometry, microwave radiometry can give an insight into the vertical structure of a precipitating cloud, while, with respect to microwave radar, it can be considered a relatively low-cost technology while providing considerably larger swaths. On the other hand, infrared radiometers have a much better sensitivity to cloud coverage and can be embarked on geosynchronous satellites, while radars provide a detailed and direct information on the vertical structure of precipitating clouds. Efforts towards a synergy between microwave radiometers and other spaceborne sensors thave represented a new trend in the last decade, as briefly described in this report, especially after the launch of TRMM. The emerging issues of near-real time monitoring of rainfall on a global-scale has put in evidence the unavoidable limit of spaceborne microwave radiometers, which is their scarce time sampling due to their installation aboard LEO platforms. This drawback is strictly related with the potential use of microwave radiometry and its products within assimilation procedures of general circulation models for weather forecasting and within natural-hazard alarm systems. Starting from these

limitations and from the solutions proposed so far, we can look through the future scenario emphasizing the following considerations:
- The Global Precipitation Mission will address the temporal sampling issue in a systematic way; nevertheless, due to the high variability of precipitation, it will be necessary to keep exploiting the potential of integrating the accurate microwave estimates of rainfall with the high temporal sampling provided by the infrared radiometers aboard next generation of geostationary meteorological satellites, such as MSG;
- Any synergetic approach between different sensors and platforms will have to tackle the problems due to the different observation geometries – even for sensors aboard the same platform, as in TRMM or the GPM core satellite – as well as to the time-delay between the observations of a rapidly evolving process (such as convective precipitation) from different platforms..
- Statistical techniques can be successfully used for integrating data acquired from different sensors, even though other approaches, like those based on neural networks, could accept in a much easier way a set of non-homogeneous input data. The choice of a combination technique should take into account the problem of time efficiency, especially if dealing with near-real-time monitoring of rainfall.

7. REFERENCES

Adler R.F., G.J. Huffman, and P.R. Keehn, 1994: Global tropical rain estimates from microwave-adjusted geosynchronous IR data. *Rem. Sens. Reviews*, **11**, 125-152.

Atlas D., D. Rosenfeld, D.B. Wolff: 1990. Climatologically tuned reflectivity-rainrate relations and links to area-time integrals. *J. Appl. Meteor.*, **3**, 1120-1135.

Bauer P. and P. Schluessel, 1993: Rainfall, total water, ice water and water vapor over the sea from polarized microwave simulations and SSM/I data. *J. Geophys. Research*, **98** (**D11**), 737-759.

Bauer P. and N.C. Grody, 1995: The potential of combining SSM/I and SSM/T2 measurements to improve the identification of snow cover and precipitation. *IEEE Trans. Geosci. Rem. Sens.*, **33**, 252-261.

Bauer P., A. Khain, I. Sednev, R. Meneghini, C. Kummerow, F.S. Marzano, and J.P.V. Poiares Baptista, 2000: Combined cloud-microwave radiative transfer modeling of stratiform rainfall. *J. Atmos. Sci.*, **57**, 1082-1104.

Bauer P., 2001: Over-ocean rainfall retrieval from multisensor data of the tropical rainfall measuring mission. Part I: Design and evaluation of inversion databases. *J. Atmos. Oceanic Technol.*, **18**, 1315-1330.

Conner M.D. and G.W. Petty, 1998: Validation and intercomparison of SSM/I rain-rate retrieval methods over the continental United States. *J. Appl. Meteor.*, **37**, 679-700.

Crosson W.L., C.E. Duchon, R. Raghavan, and S.J. Goodman, 1996: Assessment of rainfall estimates using a standard Z-R relationship and the probability matching method applied to composite radar data in central Florida. *J. Appl. Meteor.*, **35**, 1203-1219.

d'Auria G., F.S. Marzano, N. Pierdicca, R. Pinna Nossai, P. Basili, and P. Ciotti, 1998: Remotely sensing cloud properties from microwave radiometric observations by using a modeled cloud database. *Radio Sci.*, **33**, 369-392.

Di Michele S., F.S. Marzano, A. Mugnai, A. Tassa, and J.P.V. Poiares Baptista, 2001: Physically-based statistical integration of TRMM microwave measurements for precipitation profiling. *Radio Sci.*, in press.

Durden S.L., E. Im, F.K. Li, W. Ricketts, A. Tanner, and W. Wilson, 1994: ARMAR: an airborne rain-mapping radar. *J. Atmos. Oceanic Technol.*, **11**, 727-737.

Evans K.F., J. Turk, J. Wong, and T.L. Stephens, 1995: A Bayesian approach to microwave precipitation profile retrieval. *J. Appl. Meteor.*, **34**, 260-279.

Ferraro R.R. and G. F. Marks, 1995: The development of SSM/I rain-rate retrieval algorithms using ground-based radar measurements, *J. Atmos. Oceanic Technol.*, **12**, 755-772.

Haddad Z.S., S.L. Durden, and E. Im, 1996: Stochastic filtering of rain profiles using radar, surface-referenced radar, or combined radar-radiometer measurements. *J. Appl. Meteor.*, **35**, 229-242.

Hong Y.C., C.D. Kummerow, and W.S. Olson, 1999: Separation of convective and stratiform precipitation using microwave brightness temperature, *J. Appl. Meteor.*, **38**, 1195-1213.

Houze R.A., 1981: Structures of atmospheric precipitation systems: a global survey. *Radio Sci.*, **16**, 671-689.

Kummerow C. and L. Giglio, 1994: A passive microwave technique for estimating rainfall and vertical structure information from space. Part I: algorithm description. *J. Appl. Meteor.*, **33**, 3-18.

Kummerow C.D., W.S. Olson, and L. Giglio, 1996: A simplified scheme for obtaining precipitation and vertical hydrometeor profiles from passive microwave sensors", *IEEE Trans. Geosci. Remote Sensing.*, **34**, 12313-1232.

Kummerow C., W. Barnes, T. Kozu, J. Shiue, and J. Simpson, 1998: The Tropical Rainfall Measuring Mission (TRMM) sensor package, *J. Atmos. Oceanic Technol.*, **15**, 809-817.

Kummerow C., J. Simpson, O. Thiele, W. Barnes, A.T.C. Chang, E. Stocker, R.F. Adler, A. Hou, R. Kakar, F. Wentz, P. Ashcroft, T. Kozu, Y. Hong, K. Okamoto, T. Iguchi, H. Kuroiwa, E. Im, Z. Haddad, G. Huffman, B. Ferrier, W.S. Olson, E. Zipser, E.A. Smith, T.T. Wilheit, G. North, T. Krishnamurti, and K. Nakamura, 2000: The status of the Tropical Rainfall Measuring Mission (TRMM) after two years in orbit. *J. Appl. Meteor.*, **39**, 1965-1982.

Kummerow, C., D.B. Shin, Y. Hong, W.S. Olson, S. Yang, R.F. Adler, J. McCollum, R. Ferraro, G. Petty, and T.T. Wilheit, 2001: The evolution of Goddard Profiling Algorithm (GPROF) for rainfall estimation from passive microwave sensors. *J. Appl. Meteor.*, **40**, 1801-1820.

Levizzani V., F. Porcù, F.S. Marzano, A. Mugnai, E.A. Smith, and F. Prodi, 1996: Investigating a SSM/I microwave algorithm to calibrate METEOSAT infrared instantaneous rainrate estimates. *Meteorol. Appl.*, **3**, 5-17.

Marécal, V. and J.-F. Mahfouf, 2002: Four dimensional variational assimilation of total column water vapor in rainy areas. *Mon. Wea. Rev.*, **130**, 43-58.

Marzano F.S., A. Mugnai, E.A. Smith, X. Xiang, J. Turk, and J. Vivekanandanan, 1994: Active and passive remote sensing of precipitating storms during CaPE. Part II: Intercomparison of precipitation retrievals from AMPR radiometer and CP-2 radar. *Meteorol. Atmos. Phys.*, **10**, 29-54.

Marzano F.S., A. Mugnai, G. Panegrossi, N. Pierdicca, E.A. Smith, and J. Turk, 1999: Bayesian estimation of precipitating cloud parameters from combined measurements of spaceborne microwave radiometer and radar. *IEEE Trans. Geosci. Remote Sensing*, **37**, 596-613.

Marzano F.S., J. Turk, P. Ciotti, S. Di Michele, and N. Pierdicca, 2001: Potential of combined spaceborne microwave and infrared radiometry for near real-time rainfall attenuation monitoring along earth-satellite links. *Int. J. Satell. Commun.*, **19**, 385-412.

Marzano F.S. and P. Bauer, 2001: Sensitivity analysis of airborne microwave retrieval of stratiform precipitation to the melting layer parameterization. *IEEE Trans. Geosci. Remote Sensing.*, **39**, 75-91.

McGaughey G., E.J. Zipser, R.W. Spencer, and R. Hood, 1996: High-resolution passive microwave observations of convective systems over the tropical pacific ocean. *J. Appl. Meteor.*, **35**, 1921-1947.

Meneghini R., H. Kumagai, J.R. Wang, T. Iguchi, and T. Kozu, 1997: Microphysical retrievals over stratiform rain using measurements from an airborne dual-wavelength radar-radiometer. *IEEE Trans. Geosci. Remote Sensing*, **35**, 487-506.

Miller S.W., P.A. Arkin, and R. Joyce, 2001: A combined microwave/infrared rain rate algorithm. *Int. J. Remote Sensing*, **22**, 3285-3307.

Mugnai A., H.J. Cooper, E.A. Smith, and G.J. Tripoli, 1990: Simulation of microwave brightness temperatures of an evolving hail storm at SSM/I frequencies. *Bull. Amer. Meteor. Soc.*, **71**, 2-13.

Mugnai A., E.A. Smith and G.J. Tripoli, 1993: Foundations for statistical-physical precipitation retrieval from passive microwave satellite measurements. Part II: Emission source and generalized weighting function properties of a time-dependent cloud-radiation model. *J. Appl. Meteor.*, **32**, 17-39, 1993.

Mugnai A., S. Di Michele, F.S. Marzano, and A. Tassa, 2001: Cloud-model based Bayesian techniques for precipitation profile retrieval from TRMM microwave sensors", *ECMWF/EuroTRMM Workshop on Assimilation of Clouds and Precipitation, European Centre for Medium-Range Weather Forecasts*, Reading (UK), 323-345.

Mugnai A. and J. Testud, 2002: EGPM: European contribution to the Global Precipitation Mission", *Proposal in response to The Second Call for Proposals for ESA Earth Explorer Opportunity Missions*.

Negri A.J., R. Adler, and P.J. Wentzel, 1984: Rain estimation from satellites: an examination of the Griffith-Woodley technique. *J. Clim. Appl. Meteor.*, **23**, 102-116.

Olson W.S., C.D. Kummerow, G.M. Heymsfield, and L. Giglio, 1996: A method for combined passive-active microwave retrievals of cloud and precipitation parameters. *J. Appl. Meteor.*, **35**, 1763-1789.

Panegrossi G., S. Dietrich, F.S. Marzano, A. Mugnai, E.A. Smith, X. Xiang, G. J. Tripoli, P.K. Wang, and J.P.V. Poiares Baptista, 1998: Use of cloud model microphysics for passive microwave-based precipitation retrieval: significance of consistancy between model and measurement manifolds. *J. Atmos. Sci.*, **55**, 1644-1673.

Petty G., 1995:The status of satellite-based rainfall estimation over land. *Rem. Sens. of Environ.*, **51**, 125-137.

Pierdicca N., F.S. Marzano, G. d'Auria, P. Basili, P. Ciotti, and A. Mugnai, 1996: Precipitation retrieval from spaceborne microwave radiometers based on maximum *a posteriori* probability estimation. *IEEE Geosci. Remote Sensing*, **34**, 831-846.

Simpson J.R., R.F. Adler, and G.R. North, 1988: A proposed Tropical Rainfall Measuring Mission (TRMM) satellite. *Bull. Amer. Meteor. Soc.*, **69**, 278-295.

Skofronick-Jackson G.M. and A.J. Gasiewskii, 1995: Nonlinear statistical precipitation retrievals using simulated passive microwave imagery. *IEEE Trans. Geosci. Remote Sensing*, **33**, 957-970.

Smith E.A., A. Mugnai, H.J. Cooper, G.J. Tripoli, and X. Xiang, 1992: Foundations for statistical-physical precipitation retrieval from passive microwave satellite measurements. Part I: Brightness-temperature properties of a time-dependent cloud-radiation model. *J. Appl. Meteor.*, **31**, 506-531.

Smith E.A., X. Xiang, A. Mugnai, and G.J. Tripoli, 1994a: Design of an inversion-based precipitation profile retrieval algorithm using an explicit cloud model for initial guess microphysics", *Meteorol. Atmos. Phys.*, **54**, 53-78.

Smith E.A., C. Kummerow, and A. Mugnai, 1994b: The emergence of inversion-type profile algorithms for estimation of precipitation from satellite passive microwave measurements", *Remote Sens. Rev.*, **11**, 211-242.

Smith E.A., J.E. Lamm, R. Adler, J. Alishouse, K. Aonashi, E. Barrett, P. Bauer, W. Berg, A. Chang, R. Ferraro, J. Ferriday, S. Goodman, N. Groody, C. Kidd, D. Kniveton, C.

Kummerow, G. Liu, F.S. Marzano, A. Mugnai, W. Olson, G. Petty, A. Shibata, R. Spencer, F. Wentz, T. Wilheit, and E. Zipser, 1998: Results of WetNet PIP-2 Project. *J. Atmos. Sci.*, **55**, 1483-1536.

Tao W.K., S. Lang, J. Simpson, and R. Adler, 1993: Retrieval algorithms for estimating the vertical profiles of latent heat release: their applications to TRMM. *Bull. Meteor. Soc. of Japan*, **12**, 685-700.

Tassa A., S. Di Michele, A. Mugnai, F.S. Marzano, and J.P.V. Poiares Baptista, 2002: Cloud-model based Bayesian technique for precipitation profile retrieval from TRMM Microwave Imager, *Radio Sci.*, in press.

Turk J.F., F.S. Marzano, E.A. Smith, and A. Mugnai, 1998a: Using coincident SSM/I and infrared geostationary satellite data for rapid updates of rainfall. *Proc. of IGARSS'98*, Seattle (WA, USA).

Turk J., F.S. Marzano, and A. Mugnai, 1998b: Effects of degraded sensor resolution upon passive microwave retrievals of tropical rainfall. *J. Atm. Sci.*, **55**, 1689-1705.

Turk J.F., G. Rohaly, J. Hawkins, E.A. Smith, F.S. Marzano, A. Mugnai, and V. Levizzani, 1999: Meteorological applications of precipitation estimation from combined SSM/I, TRMM and geostationary satellite data. *Microwave Radiometry and Remote Sensing of the Environment*, P. Pampaloni Ed., VSP Intern. Sci. Publisher, Utrecht (The Netherlands), 353-363.

Turk, J., E. E. Ebert, H.J. Oh and B.J. Sohn, 2002: Validation and Applications of a Realtime Global Precipitation Analysis. *Proc. IGARSS'02*, June 24-28, Toronto (Canada).

Vicente G.A., R.A. Scofield, and W.P. Menzel, 1998: The operational GOES infrared rainfall estimation technique. *Bull. Amer. Meteor. Soc.*, **79**, 1883-1898.

Viltard N., C. Kummerow, W.S. Olson, and Y. Hong, 2000: Combined use of the radar and radiometer of TRMM to estimate the influence of drop size distribution on rain retrievals. *J. Appl. Meteor.*, **39**, 2103-2114.

Weinman J.A., R. Meneghini and K. Nakamura, 1990: Retrieval of precipitation profiles from airborne radar and passive microwave radiometer measurements: comparison with dual-frequency radar measurements. *J. Appl. Meteor.*, **29**, 981-993.

Wilheit T.T., R. Adler, S. Avery, E. Barrett, P. Bauer, W. Berg, A. Chang, J. Ferriday, N. Grody, S. Goodman, C. Kidd, D. Kniveton, C. Kummerow, A. Mugnai, W. Olson, G. Petty, A. Shibata, and E.A. Smith, 1994: Algorithms for the retrieval of rainfall from passive microwave measurements. *Remote Sens. Rev.*, **11**, 163-194.

Yuter S.E., R.A. Houze Jr., B.F. Smull, F.D. Marks Jr., J.R. Daugherty, and S.R. Brodzik, 1995: TOGA-COARE aircraft mission summary images: an electronic atlas. *Bull. Amer. Meteor. Soc.*, **76**, 319-328.

Clouds and Rainfall by Visible-Infrared Radiometry

VINCENZO LEVIZZANI
National Research Council, Institute of Atmospheric and Climate Sciences, Bologna

Key words: Clouds, rainfall, microphysics, nowcasting, visible, infrared.

Abstract: A precise understanding of the Earth's clouds and precipitation processes represents the natural key for a weather forecasting strategy down to the nowcasting space and time scales where cloud processes play a role and must be accurately parameterized into models. Clouds are also a governing factor in the quantification of the Earth's radiation budget, which is in turn crucial for climate and global change issues. A wide variety of satellite sensors is now in orbit or about to be launched with an unprecedented spectral width, sensitivity and space-time resolution. In the following we will examine just a few of the new possibilities offered to science and operational meteorology and climatology by the new generation VIS/IR sensors.

1.   CLOUD STRUCTURE

Deep convective clouds play a substantial role in the energy cycles of the planet through dynamic, radiative and microphysical processes. Several major question marks on the water cycle stand unanswered, such as the energy cycle of tropical storms (latent heat release), the troposphere-stratosphere water vapor exchange, the global and local distribution of precipitation, and the role of clouds in the energy cycle.

Deep convection overturns the troposphere injecting water vapor and other gases through the tropopause in the lower stratosphere (Kley et al., 1982). Hydration and dehydration effects follow, the latter due to the entrainment of warm and dry stratospheric air into the turret and the anvil and ice crystal fallout (Danielsen, 1993). Cirrus cloud formation at the top of deep convection are thus very much relevant for the water and radiation balance quantification. Tjemkes et al. (1997) have observed considerable amounts of water vapor on top of deep convective clouds using METEOSAT's water vapor (WV) channel. Measurements and numerical simulations on the composition of cumulonimbus tops have been conducted, among others, by Adler and Mack (1986), Foster et al. (1995), Fulton and Heymsfield (1991), Heymsfield and Blackmer (1988), Heymsfield et al. (1983, 1991), Knollenberg et al. (1993), Mack et al. (1983), McFarquhar and Heymsfield (1996), McFarquhar et al. (1999), McKee and Klehr (1978), Negri (1982), Pilewskie and Twomey (1987), Platnick and Twomey (1994).

The understanding of cirrus microphysical and radiative properties has considerably increased in recent years in the wake of specific experiments. Key results are those by Arnott et al. (1994), Francis et al. (1999), Gayet et al. (1996), Gothe Betancor and Graßl (1993), Heintzenberg et al. (1996), Heymsfield and McFarquhar (1996), Hutchison (1999), Kinne et al. (1992), Knap et al. (1999), Masuda and Takashima (1990), Platt et al. (1989), Ramaswamy and Detwiler (1986), Zhang et al. (1994). Wiegner et al. (1998) have tackled the problem of cirrus cloud effects in the METEOSAT Second Generation (MSG) Spinning Enhanced Visible and InfraRed Imager (SEVIRI). Cirrus radiative properties in the infrared (IR) can be found in Stephens (1980).

The retrieval of cloud optical and microphysical properties from satellite sensors is a problem yet to be fully solved. Pioneering work was carried out by Slingo and Schecker (1982) and Arking and Childs (1985). Kleespies (1995) used the 3.9 μm window for retrieving properties of marine stratiform clouds. Nakajima and King (1990) and Nakajima and Nakajima (1995) have conceived a method for the retrieval of optical thickness and effective particle radius (r_e) from reflected solar radiation measurements and applied it to the Advanced Very High Resolution Radiometer (AVHRR) observations. A recent application of the method to low-level marine clouds can be found in Kuji et al. (2000). Ou et al. (1993) and Takano and Liou (1989) have done other fundamental studies on the subject. The use of the next MSG SEVIRI instrument for the derivation of cloud properties at the geostationary scale has been extensively examined by Watts et al. (1998).

These retrievals require a precise knowledge of 1) ice crystal properties in natural clouds (Auer and Veal, 1970), 2) ice optical properties (Warren, 1984), and 3) the lifetime of natural ice crystals in the upper troposphere and stratosphere (Peter and Baker, 1996). Parameterizations are also crucial for effective simulations of the radiative behavior of cloud top (Espinoza and Harshvardhan, 1996; Fu and Liou, 1993; Minnis et al., 1998) and for climate models (Hu and Stamnes, 1993).

The role of small ice crystals in determining the radiation field on top of tropical anvils was examined by Zender and Kiehl (1994). Note that the characterization of ice and water clouds in terms of their microphysical content, r_e above all, is still a foremost research topic. Important contributions came from Baran et al. (1999), Karlsson and Liljas (1990), Liljas (1986), Parol et al. (1991), Rao et al. (1995), Rosenfeld and Gutman (1994), Rosenfeld and Lensky (1998), Saunders and Kriebel (1988). The new polarized instrument POLDER (POLarization and Directionality of Earth Reflectances) has provided unprecedented perspectives (Doutriaux-Boucher et al. 2000; Labonnote et al., 2000).

The non-sphericity and irregularity of ice cloud particles practically prevent us to use a single self-consistent theory, such as Mie's, for the rigorous definition of the scattering properties (Fu and Liou, 1993; Welch et al., 1980; Wendling et al., 1979). Computations have been performed using different ice crystal shapes and they show that scattering properties substantially vary according to shape and size of the ice particles (Asano and Sato, 1980; Macke et al., 1996, 1998; Takano and Liou, 1989). The sensitivity of cloud radiative forcing, solar albedo and IR emittance to the ice crystal spectrum and shape was studied by Zhang et al. (1999) using hexagonal columns and random polycrystals.

These uncertainties and the non-exhaustive characterization of cloud features only in terms of optical depth and ice crystal size prompt for radiative transfer modeling to simulates the interactions between clouds and radiation field. These studies are necessary for an effective satellite detection of precipitating clouds and an improvement of the existing rainfall estimation algorithms (Lensky and Rosenfeld, 1997). Rosenfeld (1999, 2000a, b) has proved their importance using Tropical Rainfall Measuring Mission (TRMM) data (Kummerow et al., 1998) and flight penetrations for the detection of the influence of natural and industrial aerosols on cloud and precipitation formation.

The retrieval of cloud top properties from a geostationary orbit is required for wider meteorological applications. Turk et al. (1998) have used the new channels in the near IR (NIR) of the Geostationary Operational Environmental Satellite (GOES-8/9) for the microphysical characterization of fog and stratus clouds. The characteristics of the forthcoming MSG SEVIRI (Schmetz et al., 2002) are very promising in this respect since it will take images of the entire Earth disc in 12 spectral channels, between 0.5 and 13.4 µm. The sampling distance at the sub-satellite point will be 3 km, except for the High Resolution Visible (HRV) channel (0.5-0.9 µm) that will sample at 1×1 km^2 resolution at sub-satellite point. The nominal repeat cycle foresees a full disc scan every 15 min. Such features will allow for a characterization of complex cloud structures by following their time evolution and improving over the insufficient coverage of polar orbiters. Greenwald and Christopher (2000) have already analyzed the comparable

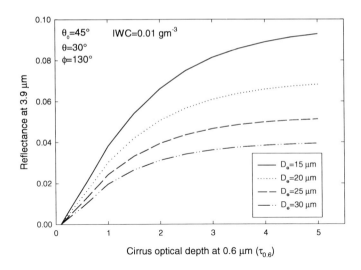

Figure 1. Sensitivity of the reflectance in the 3.9 μm channel of MSG SEVIRI to ice crystal mean effective size (D_e) as a function of the optical depth at 0.6 μm of a thin cirrus layer on top of a deep convective storm.

capabilities of the GOES I-M radiometer to provide quantitative information on the bulk microphysical properties of boundary layer stratiform clouds.

The simulations at 3.9 μm are particularly interesting for the investigations on the behavior of hydrometeors on top of deep convective clouds. Levizzani and Setvák (1996) found small ice crystal and water vapor features referred to as *plumes* because of their shape resembling a plume or a fan emanating from a relatively small spot on the cloud top. Their presence was detected by analyzing multispectral, high-resolution imagery from the AVHRR (Setvák and Doswell, 1991), especially the significant increase of the AVHRR 3.7 μm channel cloud-top reflectivity. Successive GOES-8 observations confirmed that plumes are a relatively frequent feature of deep convection. This suggests they are most likely composed of ice crystals whose size is of the order of the wavelength of channel 3. Recent high-resolution simulations with the Wisconsin Dynamical/Microphysical Model (WISCDYMM), a 3D quasi-compressible, time-dependent, non-hydrostatic cloud model, attribute the initial mechanism of plume formation to the rise of moist air blobs on top of the overshooting dome (Wang, 2001). In view of the availability of SEVIRI data, simulations of the reflectance and brightness temperature (T_B) fields in the channels centered at 0.8, 1.6, 3.9, 10.8, 12.0 μm were done by Melani et al. (2000) for a deep convective cloud with a superimposed cirrus layer using the radiative transfer model STREAMER

(Key, 1999; Key and Schweiger, 1998). The results show the sensitivity of the SEVIRI channels to ice crystal size and cloud optical depth (Fig. 1).

2. PRECIPITATION PHYSICS

2.1 VIS and thermal IR techniques

A complete overview of the early work and physical premises of visible (VIS) and thermal IR (10.5–12.5 µm) satellite rainfall estimation techniques is provided by Barrett and Martin (1981), Kidder and Vonder Haar (1995), while Levizzani et al. (2002) provide the most recent overview of the field. Levizzani et al. (2001) offer a relatively up-to-date overview of the field from a geostationary perspective. Following Barrett and Martin's (1981) classification the rainfall estimation methods can be divided in the following categories: cloud-indexing, bi-spectral, life history and cloud model-based.

Cloud indexing techniques assign a rainrate level to each cloud type identified in the satellite imagery. Arkin (1979) developed the GOES Precipitation Index (GPI) on the basis of a high correlation between radar-estimated precipitation and fraction of the area covered by pixels colder than 235 K in the IR (Arkin and Meisner, 1987). The scheme assigns these areas a constant rainrate of 3 mm h^{-1}, which is appropriate for tropical precipitation over 2.5°×2.5° areas. The GPI is a standard for climatological rainfall analysis (Arkin and Janowiak, 1991) and is regularly applied and archived for climatological studies. The World Climate Research Programme (WCRP) with its Global Precipitation Climatology Project (GPCP) has started using GPI and other global techniques for estimating precipitation for periods from five days to one month (Huffman et al., 1997, 2001). For a review on climatic-scale satellite rainfall estimates see Arkin and Ardanuy (1989).

Turpeinen et al. (1987) adapted the GPI to the current METEOSAT satellites using a set of African raingauges and examining the role of the upper tropospheric humidity (UTH) in the vicinity of convective clouds as additional predicting parameter, although results were not conclusive. The GPI was also reconstructed from METEOSAT's Climatological Data Set (CDS) comparing favorably with the actual index derived from IR counts for pixels whose temperature was lower than 235 K (Kerrache and Schmetz, 1988). Ba and Nicholson (1998) have recently analyzed the convective activity and its relationship to rainfall over East Africa using an index of convection based on METEOSAT IR data and rainfall measurements:

positive correlations are found justifying the use of the index for annual and area-averaged monthly rainfall estimates.

A family of cloud indexing algorithms was developed at the University of Bristol, originally for polar-orbiting satellites of the National Oceanic and Atmospheric Administration (NOAA) and adapted to geostationary satellite imagery. "Rain days" are identified from the occurrence of IR T_B below a threshold at a given location and combined in spatially variable means to produce rainfall estimations for periods of 10 or more days. The most recent version of the technique uses variable IR rain/no-rain threshold temperatures (Todd et al., 1995; 1999).

Bi-spectral methods are based on the very simple relationship between cold and bright clouds and high probability of precipitation, a characteristics of cumulonimbus. Lower probabilities are associated to cold but dull (thin cirrus) or bright but warm (stratus) clouds. The RAINSAT technique (Bellon et al., 1980) screens out cold but not highly reflective clouds or those that are highly reflective but have a relatively warm top. The number of false alarms of the pure IR techniques is reduced. The algorithm is based on a supervised classification trained by radar to recognize precipitation from both VIS brightness and IR cloud top temperature. RAINSAT was applied to METEOSAT and optimized over the UK by Cheng and Brown (1995). Tsonis and Isaac (1985) and Tsonis (1987) developed clustering methods similar to that applied to bi-spectral cloud classification. Rain areas are determined by classifying pixel clusters in the VIS/IR histogram of the image scene and radar data are used as "ground truth" for the validation of the method.

King et al. (1995) have examined the role of VIS data in improving IR rainfall estimates. Their results show a higher correlation with validation data using VIS/IR over the IR alone for the case of warm, orographically-induced rainfall. For cold, bright clouds (e.g. cumulonimbus) the correlations are similar. Other approaches have used pattern recognition techniques applied to VIS/IR data sets. O'Sullivan et al. (1990) used brightness and textural characteristics during daytime and IR temperature patterns to estimate rainfall over a 10×10 pixel array.

Techniques that require geostationary satellite imagery are the life-history methods that rely upon a detailed analysis of the cloud's life cycle, which is particularly relevant for convective clouds. An example is the Griffith-Woodley technique (Griffith et al., 1978). A major problem arises from the presence of cirrus anvils from neighboring clouds: they often screen the cloud life cycle underneath leading to possible underestimates early in the day and overestimates towards the evening. Negri et al. (1984) have simplified the Griffith-Woodley technique eliminating cloud tracking and producing a precipitation scheme that treats each cloud as if existing only in one image. The Negri-Adler-Wetzel (NAW) scheme has been proved to perform at the same level of the Griffith-Woodley's for the tropical environment. Advantages and limitations of the technique when applied to frontal precipitation leading to flood episodes were examined by Levizzani et al. (1990). An application of the NAW for nowcasting using also radar data was shown by Porcú et al. (1999). The NOAA-NESDIS technique (Scofield, 1987; Scofield and Oliver, 1977) involves direct interaction by a meteorologist and is currently used for operational nowcasting of heavy rainfall and flash floods (Vicente and Scofield, 1996). These methods perform reasonably for deep convective storms while contradictory results arise for stratiform systems or weak convection (Amorati et al., 1999).

Cloud model techniques aim at a quantitative improvement of the retrieval from the overall better physical description of the rain formation processes. Gruber (1973) first introduced a cumulus convection parameterization to relate fractional cloud cover to rainrate. Wylie (1979) used a cloud model to adjust calibration coefficients. A one-dimensional cloud model relates cloud top temperature to rainrate and rain area in the Convective Stratiform Technique (CST) (Adler and Negri, 1988). Local IR temperature minima are sought and screened to eliminate thin, non-precipitating cirrus. Precipitation is assigned to convective areas by means of the cloud model. To every other element colder than the stratiform threshold a fixed stratiform rainrate of 2 mm h^{-1} is assigned. The adaptation of these methods, originally developed for tropical deep convection, to other areas of the globe or climate regimes is by no means trivial (Levizzani et al., 1990). An intercomparison of three techniques over Japan was made by Negri and Adler (1993). Bendix (2000) has applied the CST to the study of El Niño events in Ecuador and Peru.

2.2   Other IR channels

IR and NIR channels other than the thermal IR window show some potential for application to rainfall estimations. Techniques for the instantaneous delineation of convective rainfall areas using split window data were initially conceived for the AVHRR (Inoue, 1985, 1987a,b). These techniques detect non-precipitating cirrus and low-level cumulus clouds using the 10.5-11.5 and 11.5-12.5 µm channels. The information content of the split window channels partially corrects erroneous rainfall area delineation (and consequent rainfall overestimate) of simple IR techniques producing better false alarm ratios (FAR). Kurino (1997) has applied a split window technique to data from the Japanese Geostationary Meteorological Satellite (GMS). He used three parameters: the 11 µm T_B, the difference between 11 and 12 µm, and the difference between 11 and 6.7 µm. The technique was statistically characterized against digital radar data over Okinawa islands using a 6-hourly GMS-5 data set during the 1995 typhoon season (July-September) and seems suitable as a nowcasting tool on 1°×1° area average.

The 3.7 (or 3.9) µm channel, included for a long time in the AVHRR instruments, serves a variety of purposes including ice discrimination and sunglint detection. Vicente (1996) developed a simple and fast algorithm for rainfall retrieval using the 11 and 3.9 µm channels with the obvious advantage of night-time use and sensitivity to the presence of ice and water vapor. The different sensitivity of the two channels to ice and water can be used as a tool to select cloud areas associated to a higher probability of producing rain. Ba and Gruber (2001) have recently implemented the GOES Multispectral Rainfall Algorithm (GMSRA) that makes use of all five GOES channels.

2.3 Rapid imaging cycles and IR-VIS-MW synergies

Low-earth orbit MW data have been proposed by several authors for a combined use with geosynchronous IR radiances using different approaches that vary widely from the monthly to the instantaneous scale. Adler et al. (1993, 1994) have proposed the Adjusted GPI (AGPI) method that corrects the GPI monthly rainfall estimates using an adjustment factor based on MW and IR data. Xu et al. (1999a) have used MW and IR data to obtain the new Universally Adjusted GPI method providing stable estimates of monthly rainfall at various spatial scales. This latter method addresses two MW-related errors: a) sampling error caused by insufficient sampling rate, and b) measurement error of instantaneous rainrate.

The more frequent imaging repetition sequence of the last generation of geostationary meteorological satellites provides a significant advantage over the past. It is true that three-axially stabilized satellites such as GOES have the capability of interrupting the operational sequence to start on-purpose rapid scan sequences down to 30 sec intervals. All current geostationary satellites, though, are conceived to operate on a 30 min time sequence that suits the needs of synoptic meteorology while showing clear deficiencies for rainfall monitoring at both convective and frontal scales. MSG will work at 15 min time sequence thus closer to cumulonimbus cloud genesis and evolution times and to the normal scanning mode of operational weather radars.

Unequivocal needs for rapid updates of rainfall estimates over land and ocean exist for quantitative precipitation forecasting, numerical weather prediction, hydrology, and Earth-space Ka-band communications (Turk et al., 1999). Satellites are the key platform to maintain routine observations along coastlines and over the oceans. Existing SSM/I radiometers on board the Defense Meteorological Satellite Program (DMSP) satellites, the TRMM Microwave Imager (TMI), and the Advanced Microwave Sounding Unit (AMSU-B) on board NOAA satellites allow for a ever increasing coverage of precipitation systems along polar or tropical orbits.

Vicente and Anderson (1993) first recognized the need for an effective strategy combining MW-based precision and IR geostationary time repetition and coverage (see also Adler et al., 1993; Jobard and Desbois, 1994; Levizzani et al., 1996). They have tackled the calibration of IR geosynchronous data using SSM/I retrievals over the Pacific Ocean using two multilinear regressions a day that reflect into a twice-a-day calibration between MW rainfall rate and IR cloud top temperature. Liu and Curry (1992) developed a cloud classification scheme of thin and deep cloud systems and a precipitation retrieval recognizing the need to use IR data for cloud top characterization in addition to MW emission and scattering signals (Liu et al., 1995). Laing et al. (1999) studied mesoscale convective systems (MCS) in Africa deriving a relationship between SSM/I-derived precipitation and METEOSAT IR data.

Turk et al. (1999) statistically blend coincident MW and IR geostationary satellite data for rapid updates of rainfall using the probability matching theory. The technique focuses on the operational time frame with emphasis on the identification and tracking of rapidly developing rain storms. 3-hourly global rainfall analyses are also produced devoid of spatial and temporal gaps. Data are candidates for direct assimilation into the general circulation and mesoscale NWP models.

The launch of MSG SEVIRI will further applications matching MW-based rainfall retrieval algorithms with the geostationary multispectral and tracking capabilities. Multispectral cloud top analysis will be crucial together with its relationships to MW scattering signatures and cloud-to-ground lightning in order to explain the very large rainfall variability over land and oceans (Ba et al., 1998). An open problem concerns the presence within active convection systems of stratiform regions (Houze, 1997) that introduce potential problems in satellite rain estimates. Texture analysis seems a promising way to get around the problem (Lensky and Rosenfeld, 1997). Significant changes in rainfall regimes and drop size distributions from convective to stratiform type were documented by Tokay and Short (1996) by examining tropical raindrop spectra. Hong et al. (1999) have used MW emission (19 and 37 GHz) and scattering (85 GHz) for the separation of convective and stratiform precipitation areas over the oceans in the tropics. The removal of no-rain clouds in IR imagery has recently been considered using ancillary MW data by Xu et al. (1999b) who used a cloud-patch algorithm to describe cloud top structural properties taking into account the varying patch features of raining and non-raining clouds.

The Auto-Estimator technique proposed by Vicente et al. (1998) exploits IR 11 μm GOES data and radar data from the US network with applications to flash flood forecasting, numerical modeling, and operational hydrology. Rainfall is retrieved through a statistical analysis between surface radar-derived instantaneous rainfall estimates and satellite-derived IR cloud top temperatures collocated in space and time. A power law regression is computed between IR cloud top temperature and radar-derived rainfall estimates at the ground (Gagin et al., 1985). Rainfall estimates are adjusted for different moisture regimes using precipitable water and relative humidity fields from the National Center for Environmental Prediction (NCEP) Eta Model and SSM/I measurements.

3. REFERENCES

Adler, R. F., and R. A. Mack, 1986: Thunderstorm cloud top dynamics as inferred from satellite observations and a cloud top parcel model. *J. Atmos. Sci.*, **43**, 1945-1960.

Adler, R. F., and A. J. Negri, 1988: A satellite infrared technique to estimate tropical convective and stratiform rainfall. *J. Appl. Meteorol.*, **27**, 30-51.

Adler, R. F., G. J. Huffman, and P. R. Keehn, 1994: Global tropical rain estimates from microwave-adjusted geosynchronous IR data. *Remote Sens. Rev.*, **11**, 125-152.

Adler, R. F., A. J. Negri, P. R. Keehn, and I. M. Hakkarinen, 1993: Estimation of monthly rainfall over Japan and surrounding waters from a combination of low-orbit microwave and geosynchronous IR data. *J. Appl. Meteorol.*, **32**, 335-356.

Amorati, R., P. P. Alberoni, V. Levizzani, and S. Nanni, 1999: IR-based satellite and radar rainfall estimates of convective storms over northern Italy. *Meteorol. Appl.*, **7**, 1-18.

Arkin, P. A., 1979: The relationship between fractional coverage of high cloud and rainfall accumulations during GATE over the B-scale array. *Mon. Wea. Rev.*, **106**, 1153-1171.

Arkin, P. A., and B. N. Meisner, 1987: The relationship between large-scale convective rainfall and cold cloud over the Western Hemisphere during 1982-84. *Mon. Wea. Rev.*, **115**, 51-74.

Arkin, P. A., and P. E. Ardanuy, 1989: Estimating climatic-scale precipitation from space: A review. *J. Climate*, **2**, 1229-1238.

Arkin, P. A., and J. Janowiak, 1991: Analysis of the global distribution of precipitation. *Dyn. Atmos Oceans*, **16**, 5-16.

Arking, A., and J. D. Childs, 1985: Retrieval of cloud cover parameters from multispectral satellite images. *J. Climate Appl. Meteorol.*, **24**, 322-333.

Arnott, W. P., Y. Dong, J. Hallett, and M. R. Poellot, 1994: Role of small ice crystals in radiative properties of cirrus: a case study, FIRE II, November 22, 1991. *J. Geophys. Res.*, **99(D1)**, 1371-1381.

Asano, S., and M. Sato, 1980: Light scattering by randomly oriented spheroidal particles. *Appl. Opt.*, **19**, 962-974.

Auer, A. H., Jr., and D. L. Veal, 1970: The dimension of ice crystals in natural clouds. *J. Atmos. Sci.*, **27**, 919-926.

Ba, M. B., and S. E. Nicholson, 1998: Analysis of convective activity and its relationship to the rainfall over the Rift Valley lakes of East Africa during 1983-90 using the Meteosat infrared channel. *J. Appl. Meteorol.*, **37**, 1250-1264.

Ba, M. B., and A. Gruber, 2001: GOES Multispectral Rainfall Algorithm (GMSRA). *J. Appl. Meteorol.*, **40**, 1500-1514.

Ba, M. B., D. Rosenfeld, and A. Gruber, 1998: AVHRR multispectral derived cloud parameters: relationship to microwave scattering signature and to cloud-to-ground lightning. *Prepr. 9th Conf. Satellite Meteorology and Oceanography*, AMS, 408-411.

Baran, A. J., S. J. Brown, J. S. Foot, and D. L. Mitchell, 1999: Retrieval of tropical cirrus thermal optical depth, crystal size, and shape using a dual-view instrument at 3.7 and 10.8 μm. *J. Atmos. Sci.*, **56**, 92-110.

Barrett, E. C., and D. W. Martin, 1981: *The Use of Satellite Data in Rainfall Monitoring*. Academic Press, 340 pp.

Bellon, A., S. Lovejoy, and G. L. Austin, 1980: Combining satellite and radar data for the short-range forecasting of precipitation. *Mon. Wea. Rev.*, **108**, 1554-1556.

Bendix, J., 2000: Precipitation dynamics in Ecuador and Northern Peru during the 1991/92 El Niño: a remote sensing perspective. *Int. J. Remote Sensing*, **21**, 533-548.

Cheng, M., and R. Brown, 1995: Delineation of precipitation areas by correlation of METEOSAT visible and infrared data with radar data. *Mon. Wea. Rev.*, **123**, 2743-2757.

Danielsen, E. F., 1993: In situ evidence of rapid, vertical, irreversible transport of lower tropospheric air into the lower tropical stratosphere by convective cloud turrets and by larger-scale upwelling in tropical cyclones. *J. Geophys. Res.*, **98(D5)**, 8665-8681.

Doutriaux-Boucher, M., J.-C. Buriez, G. Brogniez, L. C. Labonnote, and A. J. Baran, 2000: Sensitivity of retrieved POLDER directional cloud optical thickness to various ice particle models. *Geophys. Res. Lett.*, **27**, 109-112.

Espinoza, R. C., Jr., and Harshvardhan, 1996: Parameterization of solar near-infrared radiative properties of cloudy layers. *J. Atmos. Sci.*, **53**, 1559-1568.

Foster, T., W. P. Arnott, J. Hallett, and R. Pueschel, 1995: Measurements of ice particles in tropical cirrus anvils: importance in radiation balance. *Prepr. Conf. Cloud Phys.*, AMS, 419-424.

Francis, P. N., J. S. Foot, and A. J. Baran, 1999: Aircraft measurements of the solar and infrared radiative properties of cirrus and their dependence on ice crystal shape. *J. Geophys. Res.*, **104(D24)**, 31685-31695.

Fu, Q., and K.-N. Liou, 1993: Parameterization of the radiative properties of cirrus clouds. *J. Atmos. Sci.*, **50**, 2008-2025.

Fulton, R., and G. M. Heymsfield, 1991: Microphysical and radiative characteristics of convective clouds during COHMEX. *J. Appl. Meteorol.*, **30**, 98-116.

Gagin, A., D. Rosenfeld, and R. E. Lopez, 1985: The relationship between height and precipitation characteristics of summertime convective cells in South Florida. *J. Atmos. Sci.*, **42**, 84-94.

Gayet, J.-F., G. Febvre, G. Brogniez, H. Chepfer, W. Renger, and P. Wendling, 1996: Microphysical and optical properties of cirrus and contrails: cloud field study on 13 October 1989. *J. Atmos. Sci.*, **53**, 126-138.

Gothe Betancor, M., and H. Graßl, 1993: Satellite remote sensing of the optical depth and mean crystal size of thin cirrus and contrails. *Theor. Appl. Climatol.*, **48**, 101-113.

Greenwald, T. J., and S. A. Christopher, 2000: The GOES I-M imager: new tools for studying the microphysical of boundary layer stratiform clouds. *Bull. Amer. Meteor. Soc.*, **81**, 2607-2619.

Griffith, C. G., W. L. Woodley, P. G. Grube, D. W. Martin, J. Stout, and D. N. Sikdar, 1978: Rain estimation from geosynchronous satellite imagery - Visible and infrared studies. *Mon. Wea. Rev.*, **106**, 1153-1171.

Gruber, A., 1973: Estimating rainfall in regions of active convection. *J. Appl. Meteorol.*, **12**, 110-118.

Heintzenberg, J., Y. Fouquart, A. Heymsfield, J. Ström, and G. Brogniez, 1996: Interactions of radiation and microphysics in cirrus. In *Clouds, Chemistry and Climate*, P. J. Crutzen and V. Ramanathan, Ed., *NATO ASI Series*, **135**, Springer, 29-55.

Heymsfield, A. J., and G. McFarquhar, 1996: High albedos of cirrus in the tropical pacific warm pool: microphysical interpretations from CEPEX and from Kwajalein, Marshall Islands. *J. Atmos. Sci.*, **53**, 2424-2451.

Heymsfield, G. M., and R. H. Blackmer, Jr., 1988: Satellite-observed characteristics of Midwest severe thunderstorm anvils. *Mon. Wea. Rev.*, **116**, 2200-2224.

Heymsfield, G. M., R. H. Blackmer, Jr., and S. Schotz, 1983: Upper-level structure of Oklahoma tornadic storms on 2 May 1979. I: Radar and satellite observations. *J. Atmos. Sci.*, **40**, 1741-1755.

Heymsfield, G. M., R. Fulton, and J. D. Spinhirne, 1991: Aircraft overflight measurements of Midwest severe storms: implications on geosynchronous satellite interpretations. *Mon. Wea. Rev.*, **119**, 436-456.

Hong, Y., C. D. Kummerow, and W. S. Olson, 1999: Separation of convective and stratiform precipitation using microwave brightness temperature. *J. Appl. Meteorol.*, **38**, 1195-1213.

Houze, R. A., 1997: Stratiform precipitation in regions of convection: A meteorological paradox? *Bull. Am. Meteorol. Soc.*, **78**, 2179-2196.

Hu, Y. X. and K. Stamnes, 1993: An accurate parameterization of the radiative properties of water clouds suitable for use in climate models. *J. Climate*, **6**, 70-83.

Hutchison, K. D., 1999: Applications of AVHRR/3 imagery for the improved detection of thin cirrus clouds and specifications of cloud-top phase. *J. Atmos. Oceanic Technol.*, **16**, 1885-1899.

Huffman, G. J., R. F. Adler, P. Arkin, A. Chang, R. Ferraro, A. Gruber, J. Janowiak, A. McNab, B. Rudolf, and U. Schneider, 1997: The Global Precipitation Climatology Project (GPCP) combined precipitation dataset. *Bull. Am. Meteorol. Soc.*, **78**, 5-20.

Huffman, G. J., R. F. Adler, M. M. Morrissey, D. T. Bolvin, S. Curtis, R. Joyce, B. McGavock, and J. Susskind, 2001: Global precipitation at one-degree daily resolution from multisatellite observations. *J. Hydrometeorol.*, **2**, 36-50.

Inoue, T., 1985: On the temperature and effective emissivity determination of semi-transparent cirrus clouds by bi-spectral measurements in the 10 μm window region. *J. Meteorol. Soc. Japan*, **63**, 88-98.

Inoue, T., 1987a: A cloud type classification with NOAA 7 split-window measurements. *J. Geophys. Res.*, **92 D**, 3991-4000.

Inoue, T., 1987b: An instantaneous delineation of convective rainfall area using split window data of NOAA-7 AVHRR. *J. Meteorol. Soc. Japan*, **65**, 469-481.

Jobard, I., and M. Desbois, 1994: Satellite estimation of the tropical precipitation using the Meteosat and SSM/I data. *Atmos. Res.*, **34**, 285-298.

Karlsson, K. G., and E. Liljas, 1990: *The SHMI model for cloud and precipitation analysis from multispectral AVHRR data*. Swedish Meteorological and Hydrological Institute, Norrköping, SMHI PROMIS Rep., 10, 74 pp.

Kerrache, M., and J. Schmetz, 1988: A precipitation index from the ESOC climatological data set. *ESA J.*, **12**, 379-383.

Key, J., 1999: STREAMER User's Guide. *Tech. Rep. 96-01*. Dept. of Geography, Boston Univ., 90 pp., (http://stratus.ssec.wisc.edu/).

Key, J., and A. J. Schweiger, 1998: Tools for atmospheric radiative transfer: Streamer and FluxNet. *Computers & Geosciences*, **24**, 443-451.

Kidder, S. Q., and T. H. Vonder Haar, 1995: *Satellite Meteorology, An Introduction*. Academic Press, 466 pp.

King, P. W. S., W. D. Hogg, and P. A. Arkin, 1995: The role of visible data in improving satellite rainrate estimates. *J. Appl. Meteorol.*, **34**, 1608-1621.

Kinne, S., T. P. Ackerman, A. J. Heymsfield, F. P. J. Valero, K. Sassen, and J. D. Spinhirne, 1992: Cirrus microphysics and radiative transfer: cloud field study on 28 October 1986. *Mon. Wea. Rev.*, **120**, 661-684.

Kleespies, T. J., 1995: The retrieval of marine stratiform cloud properties from multiple observations in the 3.9-µm window under conditions of varying solar illumination. *J. Appl. Meteorol.*, **34**, 1512-1524.

Kley, D., A. L. Schmeltekopf, K. Kelly, R. H. Winkler, T. L. Thompson, and M. McFarland, 1982: Transport of water through the tropical tropopause. *Geophys. Res. Lett.*, **9**, 617-620.

Knap, W. H., M. Hess, P. Stammes, R. B. A. Koelemeijer, and P. D. Watts, 1999: Cirrus optical thickness and crystal size retrieval from ATSR-2 data using phase functions of imperfect hexagonal ice crystals. *J. Geophys. Res.*, **104(D24)**, 31721-31730.

Knollenberg, R. G., K. Kelly, and J. C. Wilson, 1993: Measurements of high number densities of ice crystals in the tops of tropical cumulonimbus. *J. Geophys. Res.*, **98(D5)**, 8639-8664.

Kuji, M., T. Hayasaka, N. Kikuchi, T. Nakajima, and M. Tanaka, 2000: The retrieval of effective particle radius and liquid water path of low-level marine clouds from NOAA AVHRR data. *J. Appl. Meteorol.*, **39**, 999-1016.

Kummerow, C. D., W. Barnes, T. Kozu, J. Shiue, and J. Simpson, 1998: The Tropical Rainfall Measuring Mission (TRMM) sensor package. *J. Atmos. Oceanic Technol.*, **15**, 809-817.

Kurino, T., 1997: A satellite infrared technique for estimating "deep/shallow" convective and stratiform precipitation. *Adv. Space Res.*, **19**, 511-514.

Labonnote, L. C., G. Brogniez, M. Doutriaux-Boucher, J.-C. Buriez, J.F. Gayet, and H. Chepfer, 2000: Modeling of light scattering in cirrus clouds with inhomogeneous hexagonal monocrystals. Comparison with in-situ and ADEOS-POLDER measurements. *Geophys. Res. Lett.*, **27**, 113-116.

Laing, A. G., J. M. Fritsch, and A. J. Negri, 1999: Contribution of mesoscale convective complexes to rainfall in Sahelian Africa: estimates from geostationary infrared and passive microwave data. *J. Appl. Meteorol.*, **38**, 957-964.

Lensky, I. M., and D. Rosenfeld, 1997: Estimation of precipitation area and rain intensity based on the microphysical properties retrieved from NOAA AVHRR data. *J. Appl. Meteorol.*, **36**, 234-242.

Levizzani, V., and M. Setvák, 1996: Multispectral, high-resolution satellite observations of plumes on top of convective storms. *J. Atmos. Sci.*, **53**, 361-369.

Levizzani, V., R. Amorati, and F. Meneguzzo, 2002: A review of satellite-based rainfall estimation methods. EC Project MUSIC Rep. (EVK1-CT-2000-00058), http://www.geomin.unibo.it/orgv/hydro/music/deliverables.htm, 66 pp.

Levizzani, V., F. Porcù, and F. Prodi, 1990: Operational rainfall estimation using METEOSAT infrared imagery: An application in Italy's Arno river basin – Its potential and drawbacks. *ESA J.*, **14**, 313-323.

Levizzani, V., F. Porcù, F. S. Marzano, A. Mugnai, E. A. Smith, and F. Prodi, 1996: Investigating a SSM/I microwave algorithm to calibrate METEOSAT infrared instantaneous rainrate estimates. *Meteorol. Appl.*, **3**, 5-17.

Levizzani, V., J. Schmetz, H. J. Lutz, J. Kerkmann, P. P. Alberoni, and M. Cervino, 2001: Precipitation estimations from geostationary orbit and prospects for METEOSAT Second Generation. *Meteorol. Appl.*, **8**, 23-41.

Liljas, E., 1986: *Use of the AVHRR 3.7 micrometer channel in multispectral cloud classification*. Swedish Meteorological and Hydrological Institute, Norrköping, SMHI Promis Rep., 2, 23 pp.

Liu, G., and J. A. Curry, 1992: Retrieval of precipitation from satellite microwave measurement using both emission and scattering. *J. Geophys. Res.*, **97 D**, 9,959-9974.

Liu, G., J. A. Curry, and R.-S. Sheu, 1995: Classification of clouds over the western equatorial Pacific Ocean using combined infrared and microwave satellite data. *J. Geophys. Res.*, **100 D**, 13,811-13,826.

Mack, R. A., A. F. Hasler,, and R. F. Adler, 1983: Thunderstorm cloud top observations using satellite stereoscopy. *Mon. Wea. Rev.*, **111**, 1949-1963.

Macke, A., J. Mueller, and E. Raschke, 1996: Single scattering properties of atmospheric ice crystals. *J. Atmos. Sci.*, **53**, 2813-2825.

Macke, A., P. N. Francis, G. M. McFarquhar, and S. Kinne, 1998: The role of ice particle shapes and size distributions in the single scattering properties of cirrus clouds. *J. Atmos. Sci.*, **55**, 2874-2883.

Masuda, K., and T. Takashima, 1990: Deriving cirrus information using the visible and near-IR channels of the future NOAA-AVHRR radiometer. *Remote Sens. Environ.*, **31**, 65-81.

McFarquhar, G. M., and A. J. Heymsfield, 1996: Microphysical characteristics of three anvils sampled during the central equatorial pacific experiment. *J. Atmos. Sci.*, **53**, 2401-2423.

McFarquhar, G. M., A. J. Heymsfield, A. Macke, J. Iaquinta, and S. M. Aulenbach, 1999: Use of observed ice crystal sizes and shapes to calculate mean-scattering properties and multispectral radiances: CEPEX April 4, 1993, case study. *J. Geophys. Res.*, **104(D24)**, 31763-31779.

McKee, T. B., and J. T. Klehr, 1978: Effects of cloud shape on scattered solar radiation. *Mon. Wea. Rev.*, **106**, 399-404.

Melani, S., E. Cattani, V. Levizzani, M. Cervino, and F. Torricella, 2000: Radiative effects of simulated cirrus clouds on top of a deep convective storm in METEOSAT Second Generation SEVIRI channels. Submitted to *Meteor. Atmos. Phys.*

Minnis, P., D. P. Garber, D. F. Young, R. F. Arduini, and Y. Takano, 1998: Parameterizations of reflectance and effective emittance for satellite remote sensing of cloud properties. *J. Atmos. Sci.*, **55**, 3313-3339.

Nakajima, T., and M. D. King, 1990: Determination of the optical thickness and effective particle radius of clouds from reflected solar radiation measurements. Part I: theory. *J. Atmos. Sci.*, **47**, 1878-1893.

Nakajima, T. Y., and T. Nakajima, 1995: Wide-area determination of cloud microphysical properties from NOAA AVHRR measurements for FIRE and ASTEX regions. *J. Atmos. Sci.*, **52**, 4043-4059.

Negri, A. J., 1982: Cloud-top structure of tornadic storms on 10 April 1979 from rapid scan and stereo satellite observations. *Bull. Am. Meteor. Soc.*, **63**, 1151-1159.

Negri, A. J., and R. F. Adler, 1993: An intercomparison of three satellite infrared rainfall techniques over Japan and surrounding waters. *J. Appl. Meteorol.*, **32**, 357-373.

Negri, A. J., R. F. Adler, and P. J. Wetzel, 1984: Rain estimation from satellite: An examination of the Griffith-Woodley technique. *J. Climate Appl. Meteorol.*, **23**, 102-116.

O'Sullivan, F., C. H. Wash, M. Stewart, and C. E. Motell, 1990: Rain estimation from infrared and visible GOES satellite data. *J. Appl. Meteorol.*, **29**, 209-223.

Ou, S. C., K. N. Liou, W. M. Gooch, and Y. Takano, 1993: Remote sensing of cirrus cloud parameters using advanced very-high-resolution radiometer 3.7- and 10.9-μm channels. *Appl. Optics*, **32**, 2171-2180.

Parol, F., J. C. Buriez, G. Brogniez, and Y. Fouquart, 1991: Information content of AVHRR channels 4 and 5 with respect to the effective radius of cirrus cloud particles. *J. Appl. Meteorol.*, **30**, 973-984.

Peter, T., and M. Baker, 1996: Lifetimes of ice crystals in the upper troposphere and stratosphere. In *Clouds, Chemistry and Climate*, P. J. Crutzen and V. Ramanathan, Ed., *NATO ASI Series*, **135**, Springer, 57-82..

Pilewskie, P., and S Twomey, 1987: Discrimination of ice from water in clouds by optical remote sensing . *Atmos. Res.*, **21**, 113-122.

Platnick, S., and S. Twomey, 1994: Remote sensing the susceptibility of cloud albedo to changes in drop concentration. *Atmos. Res.*, **34**, 85-98.

Platt, C. M. R., J. D. Spinhirne, and W. D. Hart, 1989: Optical and microphysical properties of a cold cirrus cloud: evidence for regions of small ice particles. *J. Geophys. Res.*, **94(D8)**, 11151-11164.

Porcù, F., M. Borga, and F. Prodi, 1999: Rainfall estimation by combining radar and infrared satellite data for nowcasting purposes. *Meteor. Appl.*, **6**, 289-300.

Ramaswamy, V., and A. Detwiler, 1986: Interdependence of radiation and microphysics in cirrus clouds. *J. Atmos. Sci.*, **43**, 2289-2301.

Rao, N. X., S. C. Ou, and K. N. Liou, 1995: Removal of the solar component in AVHRR 3.7-μm radiances for the retrieval of cirrus cloud parameters. *J. Appl. Meteorol.*, **34**, 482-499.

Rosenfeld, D., 1999: TRMM observed first direct evidence of smoke from forest fires inhibiting rainfall. *Geophys. Res. Lett.*, **26** (20), 3105-3108.

Rosenfeld, D., 2000a: Suppression of rain and snow by urban and industrial air pollution. *Science*, **287**, 1793-1796.

Rosenfeld, D., 2000b: Reduction of tropical cloudiness by soot. *Science*, **288**, 1042-1047.

Rosenfeld, D., and G. Gutman, 1994: Retrieving microphysical properties near the tops of potential rain clouds by multispectral analysis of AVHRR data. *Atmos. Res.*, **34**, 259-283.

Rosenfeld, D., and I. M. Lensky, 1998: Satellite-based insights into precipitation formation processes in continental and maritime convective clouds. *Bull. Am. Meteor. Soc.*, **79**, 2457-2476.

Saunders, R. W., and K. T. Kriebel, 1988: An improved method for detecting clear sky and cloudy radiances from AVHRR data. *Int. J. Remote Sensing*, **9**, 123-150.

Schmetz, J., P. Pili, S. A. Tjemkes, D. Just, J. Kerkmann, S. Rota, and A. Ratier, 2002: An introduction to Meteosat Second Generation (MSG). *Bull. Amer. Meteor. Soc.*, in press.

Scofield, R. A., 1987: The NESDIS operational convective precipitation technique. *Mon. Wea. Rev.*, **115**, 1773-1792.

Scofield, R. A., and V. J. Oliver, 1977: A scheme for estimating convective rainfall from satellite imagery. *NOAA Tech. Memo. NESS*, **86**, Dept. of Commerce, Washington, D.C., 47pp.

Setvák, M., and C. A. Doswell, III, 1991: The AVHRR channel 3 cloud top reflectivity of convective storms. *Mon. Wea. Rev.*, **119**, 841-847.

Slingo, A., and H. M. Schrecker, 1982: On the shortwave radiative properties of stratiform water clouds. *Q. J. R. Meteorol. Soc.*, **108**, 407-426.

Stephens, G. L., 1980: Radiative properties of cirrus clouds in the infrared region. *J. Atmos. Sci.*, **37**, 435-446.

Takano, Y., and K.-N. Liou, 1989: Solar radiative transfer in cirrus clouds. Part I: Single-scattering and optical properties of hexagonal ice crystals. *J. Atmos. Sci.*, **46**, 3-19.

Tjemkes, S. A., L. van de Berg, and J. Schmetz, 1997: Warm water vapour pixels over high clouds as observed by METEOSAT. *Contr. Atmos. Phys.*, **70**, 15-21.

Todd, M. C., E. C. Barrett, M. J. Beaumont, and J. L. Green, 1995: Satellite identification of rain days over the upper Nile river basin using an optimum infrared rain/no-rain threshold temperature model. *J. Appl. Meteorol.*, **34**, 2600-2611.

Todd, M. C., E. C. Barrett, M. J. Beaumont, and T. J. Bellerby, 1999: Estimation of daily rainfall over the upper Nile river basin using a continuously calibrated satellite infrared technique. *Meteorol. Appl.*, **6**, 201-210.

Tokay, A., and D. A. Short, 1996: Evidence from tropical raindrop spectra of the origin of rain from stratiform versus convective clouds. *J. Appl. Meteorol.*, **35**, 355-371.

Tsonis, A. A., 1987: Determining rainfall intensity and type from GOES imagery in the midlatitudes. *Remote Sens. Environ.*, **21**, 29-36.

Tsonis, A. A., and G. A. Isaac, 1985: On a new approach for instantaneous rain area delineation in the midlatitudes using GOES data. *J. Climate Appl. Meteorol.*, **24**, 1208-1218.

Turk, F. J., J. Vivekanandan, T. Lee, P. Durkee, and K. Nielsen, 1998: Derivation and applications of near-infrared cloud reflectances from GOES-8 and GOES-9. *J. Appl. Meteorol.*, **37**, 819-831.

Turk, F. J., G. D. Rohaly, J. Hawkins, E. A. Smith, F. S. Marzano, A. Mugnai, and V. Levizzani, 1999: Meteorological applications of precipitation estimation from combined SSM/I, TRMM and infrared geostationary satellite data. In: *Microwave Radiometry and Remote Sensing of the Environment*, P. Pampaloni and S. Paloscia Ed., VSP Int. Sci. Publisher, Utrecht (The Netherlands), 353-363.

Turpeinen, O. M., A. Abidi, and W. Belhouane, 1987: Determination of rainfall with the ESOC precipitation index. *Mon. Wea. Rev.*, **115**, 2699-2706.

Vicente, G. A., 1996: Algorithm for rainfall rate estimation using a combination of GOES-8 11.0 and 3.9 micron measurements. *Prepr. 8th Conf. Satellite Meteorology and Oceanography*, AMS, 274-278.

Vicente, G. A., and J. R. Anderson, 1993: Retrieval of rainfall rates from the combination of passive microwave radiometric measurements and infrared measurements. *Prepr. 20th Conf. Hurricane and Tropical Meteorol.*, AMS, 151-154.

Vicente, G. A., and R. A. Scofield, 1996: Experimental GOES-8/9 derived rainfall estimates for flash flood and hydrological applications. *Proc. The 1996 EUMETSAT Meteorological Satellite Data Users' Conf.*, EUMETSAT, 89-96.

Vicente, G. A., R. A. Scofield, and W. P. Menzel, 1998: The operational GOES infrared rainfall estimation technique. *Bull. Am. Meteorol. Soc.*, **79**, 1883-1898.

Wang, P. K., 2001: Plumes above thunderstorm anvils and their contribution to cross tropopause transport of water vapor in midlatitudes. *11th Conf. Satellite Meteorology Oceanography*, AMS, 161–164.

Warren, S. G., 1984: Optical constants of ice from ultraviolet to the microwave. *Appl. Opt.*, **23**, 1206-1225.

Watts, P.D., C. T. Mutlow, A. J. Baran, and A. M. Zavody, 1998: *Study on cloud properties derived from Meteosat Second Generation observations.* Final Rep., EUMETSAT ITT No. 97/181, 344 pp.

Welch, R. M., S. K. Cox, and W. G. Zdunkowski, 1980: Calculations of the variability of ice cloud radiative properties at selected solar wavelengths. *Appl. Opt.*, **19**, 3057-3067.

Wendling, P., R. Wendling, and H. K. Weickmann, 1979: Scattering of solar radiation by hexagonal ice crystals. *Appl. Opt.*, **18**, 2663-2671.

Wiegner, M., P. Seifert, and P. Schlüssel, 1998: Radiative effect of cirrus clouds in METEOSAT Second Generation Spinning Enhanced Visible and Infrared Imager channels. *J. Geophys. Res.*, **103(D18)**, 23217-23230.

Wylie, D. P., 1979: An application of a geostationary satellite rain estimation technique to an extra-tropical area. *J. Appl. Meteorol.*, **18**, 1640-1648.

Xu, L., X. Gao, S. Sorooshian, P. A. Arkin, and B. Imam, 1999a: A microwave infrared threshold technique to improve the GOES precipitation index. *J. Appl. Meteorol.*, **38**, 569-579.

Xu, L., S. Sorooshian, X. Gao, and H. V. Gupta, 1999b: A cloud-patch technique for identification and removal of no-rain clouds from satellite infrared imagery. *J. Appl. Meteorol.*, **38**, 1170-1181.

Zender, C. S, and J. T. Kiehl, 1994: Radiative sensitivities of tropical anvils to small ice crystals. *J. Geophys. Res.*, **99(D12)**, 25869-25880.

Zhang, Y., M. Laube, and E. Raschke, 1994: Numerical simulations of cirrus properties. *Beitr. Phys. Atmosph.*, **67**, 109-120.

Zhang, Y., A. Macke, and F. Albers, 1999: Effect of crystal size spectrum and crystal shape on stratiform cirrus radiative forcing. *Atmos. Res.*, **52**, 59-75.

Tropospheric Aerosols by Passive Radiometry

G.L. LIBERTI[1], F. CHÉRUY[2]
[1]*ISAC/CNR, Bologna, Italy;* [2]*LMD/CNRS, Paris, France*

Key words: Aerosol, remote sensing, scattering.

Abstract: Problems and solutions in the retrieval of tropospheric aerosol properties from satellite passive radiometry are described in this chapter. Because of the complexity in giving the full description of tropospheric aerosols (in terms of composition, size distribution, particle shape and vertical distribution) a preliminary section is dedicated to the definitions of aerosol components and models that are commonly used for remote sensing purposes. A relatively important issue in the satellite remote sensing of the aerosols is to separate the contribution to the measured signal due to the interaction of the electromagnetic radiation with the aerosol from the other contributions (e.g. clouds and surface). Reviewing some of the current techniques, a classification is proposed which is based on the radiation-aerosols interactions generating the observed signal. The reader should be aware that aerosol remote sensing from satellite is a field that is currently knowing an expansion, both in terms of development and availability of new sensors as well as for the development of remote sensing techniques. As a consequence, by the time this contribution will be available to the readers part of the content may need to be updated with current literature from journals.

1. INTRODUCTION

Estimation of tropospheric aerosol parameters with satellite passive remote sensing techniques is based on the measurement and inversion of electromagnetic (e.m.) radiation reaching the sensor(s) on board the platform. Such a radiation is composed of photons that have interacted with: the surface (*srf*), the non-aerosols (*na*) atmospheric components (gases and clouds) and the aerosols (*a*). In the most general case a satellite-borne radiometer produces: time (t) series of multi-spectral (λ) images (x,y) from different observation geometries (θ_o, θ_v, ϕ) of the status of e.m. radiation reaching the sensor ($I,\{I,Q,U,V\}$). As a consequence, the overall signal R at the satellite sensor can be written as:

$$R(srf, na, a) = F(t, \lambda, x, y, \theta_o, \theta_v, \phi, \{I, Q, U, V\})$$

Within this frame, the **general problem of aerosol remote sensing** may be summarized as a two step process:
1. to extract the portion of the signal containing information on the aerosols: i.e. estimate all possible non-aerosol contributions and somehow subtract them from the total signal. In practice, this leads to select observations for which the non-aerosol components (surface and atmospheric ones) are known or negligible.
2. to interpret such portion in terms of aerosol characteristics.
   Substantial differences among the techniques developed in the last two decades arise mainly from:
- the **physical process(es)** that generates the inverted signal: scattering, absorption, scattering +absorption.
- the **signature** analysed: angular, temporal, spectral, intensity and polarization. This is highly dependent on the specific satellite mission (i.e. the combination of platform and sensor(s))

In describing the remote sensing techniques for the aerosol estimation it is necessary to adopt a series of concepts and definitions both for the aerosols as well as for the radiative quantities.

Section 2 contains the basic definitions that will be employed in the rest of the chapter. In section 3, the most common factors generating non-aerosol signal are reviewed in a general way and a few examples of adopted solutions to remove them from the overall signal are given.

Section 4 contains examples of retrieval techniques that have been selected to give an overall idea of possible approaches and their specific characteristics.

For more detailed information's on the subject, the reader is referred to Kaufmann et al. (1997a) and papers of the same issue, as well as to King et al. (1999).

2. ATMOSPHERIC AEROSOLS AND RADIATIVE PROCESSES: DEFINITIONS

The concept of **atmospheric aerosols** is used to describe a **complex mixture of micron-sized** ($\approx 10^{-2} \div 10^{+2} \mu m$) **liquid or solid particles**

(except water) suspended in the air. Ideally, atmospheric aerosols are defined when, in a given air volume, the total number of aerosol particles is determined and when: size, shape and chemical composition of each single particle are known. With a similar approach the knowledge of atmospheric aerosols is clearly a problem with a huge number of unknowns. A first step in the solution of the aerosols *problem* is to reduce the number of unknowns by introducing the concept of **aerosol components** and their mixture (**aerosol models**) (e.g d'Almeida et al. 1991).

A single **aerosol component** is defined as the ensemble of particles in a given air volume originated:
- by a **source** (e.g. soil dust, sea salt particles etc.), mostly responsible for the particles chemical composition and shape;
- through a **physical process** (e.g condensation, coagulation etc.) responsible for their size distribution.

From the radiative point of view an **aerosol component** is generally defined by assuming:
- a single **shape** for all particles: mostly spherical.
- a **complex refractive index**: a variable containing the information on the chemical composition of the particles.
- a **size distribution (N(r)dr)**: in general an analytical (1 to 4 parameters) function that gives the number density of particles as a function of their radius (**r**).

An **aerosol model** is a mixture (external or internal) of aerosol components and represents the influence in a given air volume of different sources and processes responsible for the emission of the particles in the atmosphere.

Giving the above definitions, remote sensing of atmospheric aerosols is generally reduced to the determination of fewer variables:
- the **amount**, through the concepts of optical thickness τ (see Tab.1);
- and the **type**, through the selection of an aerosol component or model or a parameter, usually the Angstrom coefficient function of the model type.

In the wavelength range (0.3 to 10 μm) of interest for passive remote sensing of aerosol properties, the physical processes of interaction of the e.m. radiation with the particles are *absorption* and *scattering*. For both processes a single particle is characterized by a **cross section (C)**. The scattering process is also responsible for a redistribution, both in space as well as in term of polarization status, of the incoming radiation. Such a redistribution is expressed by the **scattering matrix (S)**.

The extinction characteristics (**absorption and scattering coefficient**) of an air volume containing a polydispersion of particles (such as an aerosol component/model) are given by the sum of the single particle characteristics.

Table 1.1 Definitions of radiometric properties of scattering/absorbing media. All quantities are wavelength dependent.

Name	Symbol	Definition	Units
Single particle scattering-absorption cross section	$C_{s/a}$	Area such that the total energy scattered/absorbed by the particle is equal to the incident radiation falling on an area of equal value	l^2
Scattering Matrix	$S(\gamma)$	$\{I,Q,U,V\} = \frac{1}{k^2 r^2} S(\gamma)\{I_o, Q_o, U_o, V_o\}$ $\{I,Q,U,V\},\{I_o,Q_o,U_o,V_o\}$: Stokes parameters of the scattered and incident light respectively, $K = 2\pi/\lambda$ wavenumber, r distance from the scatterer, γ Scattering angle	-
Phase Function	$P(\gamma)$	$S_{[1,1]}(\gamma)$	
Scattering/Absorption Coefficient	$\sigma_{s/a}$	$\int_0^\infty C_{s/a} N(r) dr$	l^{-1}
Extinction Coefficient	σ_e	$\sigma_s + \sigma_a$	l^{-1}
Optical Thickness	τ	$\int \sigma_e ds$	-
Single Scattering Albedo	ω	σ_s/σ_e	-
Angstrom coefficient	α	$-\dfrac{\ln \dfrac{\tau(\lambda_1)}{\tau(\lambda_2)}}{\ln \dfrac{\lambda_1}{\lambda_2}}$	-

More often, for consistency with the parameters needed in radiative transfer models, the equivalent parameters: **extinction coefficient** and **single scattering albedo** are used. When the extinction is computed over the geometric path of the radiation the concept of **optical thickness** (τ) is introduced.

Similarly, due to the fact that the majority of the techniques analyses the total intensity of the radiation without taking into account the polarization state of the radiation, only the term [1,1] of the scattering matrix, defined as **phase function** (P), is considered.

The spectral dependence of the extinction contains information on the characteristics of the aerosols (e.g. Shifrin, 1995; King et al., 1978), in particular on the size distribution. A common parameter used to describe such a dependency is the **Angstrom coefficient** α(Angstrom, 1964).

3. NON-AEROSOL COMPONENTS

In the most common case, as for example the analysis of scattered visible radiation, the contribution to the measured signal of the aerosol is a small portion of the overall signal to be inverted. It is intuitive, in such a case, the importance of correctly remove the non-aerosol contribution. In the following the main sources of non-aerosol contribution are presented together with examples of solution to treat them.

3.1 Clouds

Due to the relatively large optical thickness of clouds, even of thin ones, compared with common values of the atmospheric aerosols ones, the quality of aerosol products are extremely sensitive to the capability to discriminate clouds from aerosols and/or to detect even partially contaminated pixels. An excellent summary of cloud detection problems and solutions can be found in papers describing the cloud detection scheme for recent satellite missions such as: Bréon and Colzy (1999) that contains most of the concept adopted in solar-based cloud detection techniques and Ackerman et al. (1998) that concentrates mostly on the thermal infrared (TIR) ones. The main difficulties in discriminating clouds from aerosols are in the cases where their contribution to the measured signal are comparables, for examples:

- detection of thin cirrus that are relatively transparent in the TIR that cannot be identified because of their altitude. A possible solution to such problem is the analysis of radiances at wavelengths with strong absorption in the lower troposphere. Gao et Kaufmann (1995), proposed for the mission MODIS, a narrow channel near the center of the strong 1.38 μm water vapour band to detect thin cirrus as well as stratospheric aerosols;
- in presence of large loads of desert dust aerosols that is often accompanied by formation of clouds. A method to discriminate between the two uses the analysis of the local spatial variability of the radiances, based on the hypothesis that desert dust has a smoother spatial distribution than clouds;
- sub-pixel cloud detection is another very difficult issue and it is practically the main driving requirement for spatial resolution for aerosol remote sensing.

3.2 Atmospheric gas absorption

The presence of absorption, by atmospheric gases if not corrected, introduces:
– overestimation of the aerosol optical thickness;

- a spectral signature of the analysed signal with consequent errors in the aerosol model retrieval;
- a sensitivity on the vertical distribution of the aerosol..

As a consequence, satellite remote sensing of aerosol vertically integrated quantities, uses radiances measured in region of the e.m. for which the absorption from atmospheric gases is negligible (i.e. atmospheric windows).

3.3 Molecular Scattering

Analytical formulas to remove the molecular scattering contribution have been investigated by several authors (e.g. Bucholtz 1995). The main source of uncertainties are due to:
- the variations of the surface pressure, that in the specific case of satellite remote sensing is a combination of the pressure due the air mass column and of the variability due to the height of the surface within the Field of View (FOV) of the instrument;
- the variations in the air-composition, especially of the water vapor content, that are responsible mostly for the scattering matrix characteristics.

Due to the size of molecules, the molecular optical thickness has a λ^{-4} dependence. Taking into account the general $\lambda^{-1}, \lambda^{-2}$ dependence of the aerosol contribution, it is evident the advantage of using longer wavelengths to minimize the relative contribution of molecular scattering.

However, this is limited by (1) the lower sensitivity to small aerosol particles, (2) the lower energy at the detector due to the incoming solar spectrum, (3) the increasing thermal emission contribution that requires for its correction the temperature and emissivity of the emitting bodies.

3.4 Surface

Estimation of the surface contribution is, in general, the most difficult issue in the development of satellite aerosols retrieval techniques because:
- surface radiative properties (e.g. emittance, bidirectional reflectance (BDR): $\rho^{\lambda}(\theta_v, \theta_s, \phi)$, etc.) are strongly dependent on the surface cover type (vegetation, soil type).
- the surface cover type itself may be highly variable in time as well in space. The latter means the probability of inhomogeneous surface cover even within the FOV of the sensor.
- the surface radiative properties are spectrally dependent.

In addition, for most of the wavelengths and most of the surface types, the surface contribution is larger than the atmospheric one, therefore a relatively low error in the surface contribution estimation can be of the order of magnitude of the entire range of variability of the atmospheric aerosol contribution.

The exact solution of the *surface problem* would be given by either in situ measurements of BDR or exact radiative transfer computation for which still detailed input from in situ data would be necessary.

Several approaches have been proposed to solve the **surface** problem. In the following few examples are reported:

- Use of surface BDR models, mostly applied over ocean (e.g. Koepke 1984) or simplified models of land surface (e.g. Roujean 1992). Such models require generally independent information such as, for example the wind for ocean surface models or the surface type, for land surface models.
- Estimate the surface properties from time series of data from the same instrument. A similar approach is based on the hypothesis that, in general, temporal variability of surface properties is lower than for the atmospheric ones. Making some assumption on the atmospheric properties, it is possible to retrieve, from time series of satellite observations, the surface properties (e.g. Nadal and Breon 1999, Herman and Celarier 1997, Legrand et al. 1989)
- Estimate surface properties from known spectral dependency. An example is the MODIS aerosol retrieval over land (Kaufman et al. 1997) that is based on the following assumptions:
  a) surface reflectances, at different wavelengths, are correlated.
  b) aerosol scattering effect decrease with the wavelength;
  On these basis the technique searches for low reflectance pixels in the SWIR (Short Wave Infrared) channels (2134 and 3800 nm). The surface BDR's are firstly determined in the Short Wave Infrared (SWIR), where, according to (b), the atmospheric contribution is negligible. Then using empirical relationships, based on (a), SWIR BDR are used to compute the BRD values in the VIS (470 and 659 nm) where the aerosol retrieval is performed. Recently Kaufmann et al. (2000) demonstrate the validity of such approach, originally developed for relatively dark surface, for the retrieval of desert dust over desert.
- Estimate surface properties from known angular dependency. A technique to eliminate the surface contribution on the basis of the observed property that the angular dependency of BDR is independent from the wavelength in a limited wavelength range has been developed and applied to the analyses of ATSR-2 data (North et al.1999) over land. Practically the technique assumes:

$$\frac{\rho^\lambda(0^\sigma,\theta_r,\phi)}{\rho^\lambda(55^\sigma,\theta_r,\phi)} \approx \text{Costant} \quad (\lambda = 555, 659, 865 \quad and \quad 1600\ nm)$$

and apply a two stage iteration process:
1) Derive a set of land surface BDR from the top of atmosphere radiances, given an initial estimate of aerosol model and optical depth.
2) Evaluate how well these reflectances satisfy the given constraint on the angular dependence.

The iteration proceeds to find the aerosol model and optical depth that minimize the difference between estimated and expected value of the BDR.

- Analyse the state of polarization of the measured signal. Incoming solar radiation is completely unpolarized. The scattering of solar radiation from aerosols generates a significant amount of polarized light. Although the surface contributes to the polarization of natural light by means of specular reflection, its contribution to the polarized radiance, as measured by a satellite borne radiometer, is in general lower or of the same order of magnitude of the atmospheric one. Therefore, uncertainties in modelled (e.g. Cox and Munk 1954 for ocean surface) or semi-empirically (e.g. Nadal and Bréon 1999) derived surface properties have a lower relative impact on the aerosol retrieval, with respect to the intensity based one for which the surface contribution is larger than the aerosol one.
- extract simultaneously surface and aerosols properties. This approach differs from the previous one by the fact that interactions between atmosphere and surface (multiple scattering/reflections) are taken into account. This is the case, for example, of recent ocean-color algorithms (Siegel et al. 2000).

4. EXAMPLES OF AEROSOLS RETRIEVAL

4.1 Techniques in the Ultraviolet (UV)

In the UV range (340-380 nm), the most relevant contribution, in cloud-free conditions, to the measured radiance ($I(\lambda)_{meas}$) is the molecular scattering, the surface BDR being low (<0.05, except for snow) and relatively spectrally independent (Herman and Célarier 1997).

On this basis, Herman et al. (1997), using TOMS data, defined an aerosol index (AI) that compares the spectral dependence of the measured radiances $(I(340)_{meas}/I(380)_{meas})$ against the spectral dependence of radiances simulated with a radiative transfer model assuming a pure molecular (no-aerosol) atmosphere:

$$AI = -100\{\log_{10}[I(340)_{meas}/I(380)_{meas}] - \log_{10}[I(340)_{calc}/I(380)_{calc}]\}$$

In absence of aerosols, the measured radiation will approximately shows the spectral dependence of molecular scattering $(\approx \lambda^{-4})$ i.e. $AI \approx 0$

In presence of non-absorbing aerosols, the scattering will increase the intensity of the measured radiances and will introduce a lower $(\approx \lambda^{-2})$ spectral dependence. The AI will be negative or close to 0 depending on the aerosol size and surface reflectivity.

In presence of absorbing aerosols, the absorption being roughly spectrally independent in the UVB, part of signal generated by the molecular scattering, within and below the aerosol layer, will be absorbed and the AI values will result positive.

Due to the physical mechanisms generating the AI signal, its response to the atmospheric aerosol depends on the nature and the vertical distribution of the latter (Torres et al. 1998). Studies comparing AI values to independently measured aerosol optical thickness (Hsu et al. 1999, Chiappello et al. 2000) concluded that, while it is possible to define locally valid relationships in presence of dominant aerosol type (e.g. biomass burning or desert dust), it is rather difficult to define a relationship that can be applied globally. Nevertheless, TOMS derived AI maps have been widely used by the scientific community in the last few years, since, at the present time, they are the only available information of the global (land/ocean) distribution of the presence of aerosols with a relatively long record from Nov. 1978 at present.

4.2 Techniques in the Visible (VIS) and Short Wave Infrared (SWIR)

4.2.1 Introduction

As a consequence of the general size range of atmospheric aerosols and their usual values of particle density, shorter wavelengths are more sensitive to the presence of atmospheric aerosols than longer ones. Therefore, the majority of the satellite remote sensing techniques have been developed in the range of wavelength ranging from 0.4 to about 2 µm (where $VIS =\approx 0.4 \div 0.7 \mu m$ and $SWIR =\approx 0.7 \div 2.0 \mu m$) and are based on the analysis of the interaction between the radiation emitted by the sun and the atmosphere-earth system.

In order to discuss such techniques, let us introduce a simple analytical model for the computation of the signal at the satellite sensor, based on the following assumptions:

- sun as unique radiation source (i.e. no thermal emission).
- monochromatic radiation;
- plane parallel vertically homogenous atmosphere; with no absorption from atmospheric gas constituents ($\omega_m = 1$ i.e. atmospheric window) and absence of clouds (i.e. clear sky);
- negligible contribution of photons scattered and reflected for any order >1 (i.e. single scattering and single reflection),

Under such assumptions, the signal reaching the satellite sensor is the sum of three independent contributions:

$$R_t = R_s + R_m + R_a$$

where the subscripts s, m and a are used to identify the surface, the molecular and the aerosol contributions respectively.

The aerosol contribution, given in terms of normalized reflectivity (i.e. the ratio between the upwelling and the downwelling solar radiances at the top of the atmosphere), can be written as:

$$R_a = \frac{1}{4(\mu_s + \mu_v)} \cdot \omega_a \cdot P_a(\gamma) \cdot [1 - \exp(-\tau_a \cdot (1/\mu_s + 1/\mu_v))] \approx \quad (1)$$

$$\approx \underbrace{\frac{1}{4 \cdot \mu_s \cdot \mu_v}}_{\text{Observation geometry}} \cdot \underbrace{\tau_a \cdot \omega_a \cdot P_a(\gamma)}_{\text{Scatterer}} \quad (2)$$

$$\overset{\text{Amount Type}}{}$$

Where:
$\mu_v = \cos(\theta_v)$ with θ_v the viewing zenith angle
$\mu_s = \cos(\theta_s)$ with θ_s the solar zenith angle
γ is the scattering angle defined by the sun-target-satellite observation geometry as follows:

$$\gamma = \cos^{-1}[-(\cos(\theta_s) \cdot \cos(\theta_v) + \sin(\theta_s) \cdot \sin(\theta_v) \cdot \cos(\phi))]$$

where ϕ is the relative azimuth between the observer and the sun.

The approximation of eqt.1.1 by eqt.1.2 corresponds, mathematically, to the first order Taylor series expansion of the exponent assuming:

$$\tau_a \cdot (1/\mu_s + 1/\mu_v) \ll 1$$

that corresponds to low aerosol load. However, comparisons with more complex models show that this expression approximates the contribution better than eqt. 1.1 in presence of multiple scattering (Viollier et al. 1980). This is due to the fact that, while eq. 1.1 increases asymptotically for an increasing optical thickness, eqt. 1.2 compensates for the multiple scattering contribution by increasing linearly with the optical thickness.

From a physical point of view, equation 1.2 clearly shows that scattering contributions depend on the total amount of the scatterers (τ_x) times the signature of the scatterer $(\omega \cdot P_x)$.

A second level of approximation is to allow multiple interactions, i.e. multiple scattering and/or reflection. In such case, the signal at the satellite sensors will be (Chandrasekhar 1960):

$$R_t = R_{atm} + \frac{A_g}{1 - A_g \cdot r_{atm}} \tau_{atm}(\theta_s) t_{atm}(\theta_v) \qquad (3)$$

Where:
- the surface has been assumed with reflectance A_g;
- R_{atm} and r_{atm} are the upward and downward reflectances of the atmosphere, respectively;
- $t_{atm}(\theta)$ is the transmittance for a zenith angle θ.

4.2.2 Single Wavelength

Assuming both surface and molecular contribution as known and imposing an aerosol model (i.e. $P(\gamma)$ and ω), the aerosol optical thickness can be obtained from a measurement by inverting eqtn. 1.2

$$\tau_a = (R_t - R_s - R_m) \cdot \frac{4\mu_s \mu_v}{\omega_a P_a(\gamma)} \qquad (4)$$

This technique can be applied, in principle, over any surface of known BDR properties; in practice, it has been applied only over oceans (e.g. Jankowiak and Tanre ,1992; Rao et al. ,1989) mostly because of the low reflectance value of sea water and the homogeneity of the surface cover.

The main limitation of this technique is the assumption of a 'fixed' aerosol model that can introduce large uncertainties on the estimated optical thickness.

The main advantage is that, for its simplicity, it can be applied to data from operational satellite missions for which long time series of measurements are available, as for example the geostationary satellites (Moulin et al., 1997) or the satellites of the NOAA series (Husar et al., 1997).

4.2.3 Multispectral techniques

The composition and size distribution of the aerosol particles are responsible for the spectral dependence of their optical properties. As a consequence, radiances measured at a satellite sensor will contain a spectral information on the aerosol model. This can be clearly shown, with the simplified model introduced above, considering the case of 2 wavelengths λ_1 and λ_2 with both surface and non-aerosols atmospheric contribution removed. In such a case the ratio between the aerosol contribution, expressed as in eq. 1.2, will be:

$$\in(\lambda_1,\lambda_2) = \frac{\frac{1}{4\mu_o\mu_v} \cdot \tau_a(\lambda_1)\omega_a(\lambda_1)P_a(\gamma,\lambda_1)}{\frac{1}{4\mu_s\mu_v} \cdot \tau_a(\lambda_2)\omega_a(\lambda_2)P_a(\gamma,\lambda_2)} \approx \left(\frac{\lambda_1}{\lambda_2}\right)^{-\alpha} \frac{\omega_a(\lambda_1)P_a(\gamma,\lambda_1)}{\omega_a(\lambda_2)P_a(\gamma,\lambda_2)}$$

(5)

Where we assume the spectral dependence of the optical thickness described by the Angstrom coefficient. The above quantity depends only on the aerosol model and can be used to select among tabulated $\in(\lambda_1,\lambda_2)$'s for different aerosol models. Once the model is determined, it is possible to compute the optical thickness.

Multispectral retrievals of aerosols have been applied to several missions as for example: NOAA-AVHRR (Higurashi and Nakajima, 1999), TRMM-VIRS (Ignatov and Stowe, 2000), MODIS (Tanre et al., 1997, Kaufmann et al., 1997b), SeaWiFS (Gordon and Wang, 1994). The main differences among the multispectral techniques are the number of wavelengths and algorithm for the estimation of the surface contribution.

4.2.4 Analysis of the multi-angular signature

Multispectrally based techniques have limited sensitivity to aerosol type

characteristics because, for the majority of the cases, the wavelengths used are confined in a limited spectral range (about 600-1000 nm), where:
- the lower boundary is such to minimize the surface (over ocean) and molecular scattering contributions.
- the upper boundary is given by (1) sensor spectral response (for example: Si based sensors are limited to about 1000 nm), (2) the solar input energy and (3) the increasing contribution of the gas absorption.

This reduces, for most of the missions, the wavelengths to standard values of about 670, 750, 870 nm corresponding to the atmospheric windows in the 600-1000 nm range. Unfortunately, radiances within this limited wavelength range can have similar response to the aerosol models such that their information content is not sufficiently independent to separate the aerosol optical thickness from the aerosol model contribution.

Additional information can be obtained by the analysis of the multiviewing measurements in the scattering matrix, responsible for the modulation of the angular signal. It has been shown (Tonna et al., 1995; Tanre et al., 1996) that the sensitivity of scattered radiation to different class sizes depends from the scattering angle: i.e. the availability of measurements from more than a single scattering angle allows a better description of the aerosol model. In addition, the analysis of multi-angular scattered radiation is probably the only tool to determine the aerosol shape (e.g. Khan et al., 1997). This is the basis of multi-spectral/multi-angular algorithms over ocean as for example MISR (Khan et al., 1998), ATSR (Veefkind and de Leeuw 1998, Veefkind et al. 1999) and POLDER that also uses its polarization capability (Herman et al., 1997b; Deuze et al., 1997).

Besides the improved capability in the aerosol characteristics estimation, multi-angular observations are useful (see above) for the estimation of surface contribution and, over ocean, they allow to avoid specular reflection from the sun (sunglint).

4.2.5 Analysis of the polarization of the measured radiation

Besides the advantages in reducing the relative contribution of the surface, compared to the aerosol one (see above), the analyses of the polarization status of the measured radiances has been demonstrated to be efficient, especially when used with the analysis of the intensity, to help identifying the aerosol characteristics (Mishchenko and Travis, 1997). In particular, when coupled with multi-angular capabilities, as in POLDER, there is sensitivity to the refractive index and to the particle shape. Sensitivity studies and applications can be found in: Herman et al. (1997), Deuze et al. (1997).

4.2.6 Analysis of the spatio-temporal signatures in satellite images: I land contrast

These techniques are based on the hypothesis that over an observed scene the 3D spatial variability of the atmosphere is by far lower than the 2D one of the surface properties. From an intuitive point of view an aerosol-free image shows a stronger contrast compared to an hazy one (blurring effect). Considering eq. 1.3, if we assume the atmospheric contribution the same for two relatively close pixels i and j the difference in the measured signal will be:

$$\Delta R_t^{i,j} \approx \Delta A_g^{i,j} \cdot t_{atm}(\theta_s) t_{atm}(\theta_v) \qquad (6)$$

If we assume in addition that, given a time series of images, the surface characteristics are constant within the period analysed, the difference in ground reflectance ($\Delta A_g^{i,j}$) can be estimated, for example analysing a time series of images containing a relatively clear day for which optical thickness can be estimated or was measured. Then, it is possible to invert equation 1.6 and retrieve, from the transmission functions $t_{atm}(\theta)$, the aerosol optical thickness. The need of being able to estimate $\Delta A_g^{i,j}$ is the main limitation to this technique that has been applied mostly over limited areas. The contrast within an image, and the consequent sensitivity to the aerosol parameters, increases with decreasing spatial resolution (Tanre and Legrand, 1991). This also limits the applicability of such a technique. Theoretically, the blurring scale length is of the order of the effective scale height (z_{eff}) of the scattering opacity of the atmosphere, images of a surface overlaid by an atmosphere with an opacity dominated by tropospheric aerosols $(z_{eff} \approx 1-2km)$ will appear as blurred if the image resolution is less than ≈ 1 km.

4.2.7 Analysis of the spatio-temporal signatures in satellite images: II ocean-land contrast

Let us examine the atmospheric aerosol effect on the measured top of atmosphere reflectance (i.e. the difference between the surface reflectance and the TOA one):
- for a non reflecting surface, the effect will be to increase the reflectance value with increasing aerosol optical thickness because of the backscattered radiation contribution;
- for a highly reflecting surface and absorbing aerosol, the positive contribution of the aerosol backscattered radiation can be cancelled, because

of absorption and backscattering by the aerosols of the surface reflected contribution.

This effect depends on the aerosol τ_a, ω_a and the surface properties. Using such a sensitivity and analysing:

1) limited areas, for which the atmosphere can be considered homogeneous; 2) time series with at least one clear and one hazy day 3) scenes with relatively high gradient in the surface reflectance characteristics as for example coasts, it is possible to retrieve τ_a and ω_a. Although limited by the above conditions, this is the only technique that have demonstrated (Fraser and Kaufman, 1985; Kaufman ,1987) the capability to retrieve aerosol absorption properties (i.e. ω_a) whose knowledge is fundamental for all radiation applications. It has the same limitations as the ocean contrast technique. It has been applied to high (Landsat) to medium resolution (AVHRR) sensors (Kaufmann and Joseph, 1982; Kaufmann, 1987; Kaufman et al., 1990; Nakajima and Higurashi, 1997).

4.3 Techniques in the Thermal Infrared (TIR)

Top of atmosphere radiances at TIR window wavelengths are sensitive to the presence of atmospheric aerosols mostly through the absorption and emission processes rather than the scattering one. Several authors have reported on detection and quantitative estimation of atmospheric aerosols, in particular airborne desert dust and stratospheric aerosols, in the TIR range. The physical principles of TIR based aerosol remote sensing techniques can be summarized as follows:

- Detection of the temporal variability of TOA radiation over desert. These techniques have been applied mostly to data from broadband IR 10-12 μm window radiometers (e.g. Shenk and Curran, 1974; Legrand et al., 1989; Tanre and Legrand, 1991)
- Detection of distinct aerosol spectral signatures in the 3.7, 8, 11 and 12 μm narrow channels (e.g. Ackerman, 1989, 1997)
- Detection of size distribution signature, over desert surface in the 8, 11 and 12 μm narrow channels (Wald et al. 1998). At wavelengths corresponding to the atmospheric window channels, radiative characteristics of mineral dust components are such that they induce a differential spectral sensitivity of the emissivity to the size distribution. Using this property, it is possible to discriminate between coarse sand at the ground and fine suspended one.

The main problems, in currently available IR based aerosol remote sensing techniques, arise from the spectral signature of the surface as well as from the temperature and humidity profile dependence of the measured radiance. However, the TIR based aerosol retrieval is of particular help in the case of airborne desert dust over land. In this case, VIS/SWIR techniques are likely to fail due to the high reflectivity of the underlying surface, the non-sphericity of dust particles and the relatively low altitude at which desert dust is located close to the sources. Moreover, an unique advantage of TIR based aerosol remote sensing is that it can be applied, at least on limited cases, both at day and at night.

5. REFERENCES

Ackerman, S.A., K.I. Strabala, W.P. Menzel, R.A. Frey, C.C. Moeller and L.E Gumley, 1998. Discriminating clear sky from cloud with MODIS. J. Geophys. Res., 103, 32141-32158.

Ackerman, S.A., 1997. Remote sensing aerosols using satellite infrared observations. J. Geophys. Res., 102, 17069-17079.

Ackerman, S.A., 1989. Using the radiative temperature difference at 3.7 and 11micron to track dust outbreaks. Rem. Sens. of Environ., 27, 129-133.

Angstrom, A., 1964. The parameters of atmospheric turbidity. Tellus, XVI, 1, 64-75.

Bréon, F.M. and S.Colzy, 1999. Cloud detection from the spaceborne POLDER instrument and validation against surface synoptic observations. J. Appl. Meteor., 38, 777-785.

Bucholtz, A., 1995. Rayleigh scattering calculation for the terrestrial atmosphere. Appl. Opt. 34, n.15, 2765-2773.

Chandrasekhar, S., 1960. Radiative Transfer. Dover. 393 pp.

Chiappello I., Goloub P., Tanré D. et al. : Aerosol detection by TOMS and POLDER over oceanic regions. JGR, 105 (D6), 7133-7142, 2000.

Cox, C. and W. Munk, 1954. Statistics of the sea surface derived from sun glitter. J. Mar. Res., 13, 198-208, 1954.

d'Almeida, G.A., P. Koepke, and E. P. Shettle, 1991: Atmospheric Aerosols, Global Climatology and Radiative Characteristics (A.Deepak Publishing, Hampton, 1991) pp.561.

Fraser, R.S. and Y.J. Kaufman, 1985: 'The relative importance of aerosol scattering and absorption in remote sensing', IEEE J. Geosc. Rem. Sens., GE-23, 525-633.

Gao, B-C. and Y.J.Kaufman, 1995. Selection of the 1.375 \mum MODIS channel for remote sensing of cirrus clouds and stratospheric aerosols from space. J. Atmosph. Sc., 52, 4231-4237.

Gordon, H.R. and M.Wang, 1994. Retrieval of water leaving radiance and aerosol optical thickness over the oceans with SeaWiFS: Preliminary algorithm. Appl. Opt., 33, 443-452

Herman, J.R, P.K.Barthia, O.Torres, N.C.Hsu, C.J.Seftor and E.Celarier, 1997. Global distribution of UV-absorbing aerosols from Nimbus 7/TOMS data. J.Geophys.Res., 102, 16911-16922.

Herman, J.R. and E.A.Celarier, 1997. Earth surface reflectivity climatology at 340-380 nm from TOMS data. J. Geophys. Res. 102, D23, 28003-28011.

Herman, M., J.L.Deuze, C.Devaux, P.Goloub, F.M.Breon and D.Tanre, 1997b. Remote sensing of aerosols over land surfaces including polarization measurements and application to POLDER measurements. J.Geophys.Res., 102, 17039-17049.

Higurashi, A. and T. Nakajima, 1999. Developpment of a two-channel aerosol retrieval algorithm on a global scale using NOAA-AVHRR. J. Atmos. Sci., 56, 924-941.

Hsu, N.C., J.R.Herman, O.Torres, B.N.Holben, D. Tanr\'e, T.F. Eck, A. Smirnov, B. Chatenet, F lavenu, 1999, Comparisions of the TOMS aerosols index with Sunphotometer aerosols optical thickness: Results and applications, J.Geoph. Res., v.104, D6, 6269-6279

Husar, R.B., J.M.Prospero and L.L. Stowe 1997. Characterization of tropospheric aerosols over the oceans with the NOAA AVHRR optical thickness operational product. J. Geophys. Res., 102, 16889-16909.

Ignatov, A. and L. Stowe, 2000: Physical basis, premises, and self-consistency checks of aerosol retrievals from TRMM VIRS. J. Appl. Meteor., 39, n.12, 2259-2277.

Jankowiak,I., and D.Tanre 1992. Satellite climatology of Saharan dust outbreaks: method and preliminary results. J. Clim., 5, 646-656.

Kahn, R., R. West, D. McDonald,B. Rheingans and M.I. Mishchenko, 1997. Sensitivity of multi-angle remote sensing observations to aerosol sphericity J.Geoph. Res., v.102, D14, 16861-16870.

Kahn, R., P. Banerjee, D. McDonald and D. Diner, 1998. Sensitivity of multi-angle imaging to aerosols optical depth and to pure-particle size distribution and composition over ocean. J.Geoph. Res., v.103, D24, 32195-32213.

Kaufman, Y.J., A. Karnieli, and D. Tanre, 2000: Detection of dust over desert using Satellite Data in the solar wavelenght. IEEE Trans. Geosci. Remote Sensing, 38, n.1, 525-531.

Kaufman Y.J., D.Tanre, H.R.Gordon, T. Nakajima, J.Lenoble, R.Frouin, H.Russel, B.M.Herman, M.D.King and P.M.Teillet, 1997a. Passive Remote Sensing of Tropospheric aerosols and Atmospheric correction for the aerosol effect. J.Geoph. Res., v.102, D14, 16815-16830.

Kaufman,Y.J., D.Tanre, L.A.Remer, E.F.Vermote, A.Chu and B.N.Holben, 1997. Operational remote sensing of tropospheric aerosol over land from EOS/MODIS. J. Geophys. Res. 102, D14, 17051-17067.

Kaufman Y.J. and J.H.Joseph, 1992. Determination of surface albedos and aerosol extinction characteristics from satellite imagery. J. Geoph. Res., 97, 1287-1299.

Kaufman Y.J., R.S. Fraser and R.A. Ferrare, 1990 Satellite measurements of large-scale air pollution methods, J. Geoph. Res., vol. 95(D7), pp. 9895-9909.

Kaufman Y.J., 1987: Satellite Sensing of Aerosol absorption. J. Geoph. Res., 92, n.D4, 4307-4317.

King M.D., Kaufman Y.J., Tanre D., Nakajima T., 1999: Remote sensing of tropospheric aerosols from Space: Past, Present, and Future. Bull. Amer. Met. Soc., 80, 2229-2259.

King, M.D., D.M. Byrne, B.M. Herman and J.A. Reagan, 1978, Aerosol size distribution obtained by inversion of optical depth measurements, J. Atmos. Sci., 35, 2153- 2167.

Koepke, P., 1984. Effective reflectance of oceanic whitecaps, Appl. Opt., 23, 1816-1823.

Legrand, M., J.J.Nertrand, M.Desbois, L.Menenger and Y.Fouquart, 1989. The potential of infrared satellite data for the retrieval of Saharan dust optical depth over Africa. J. Clim. Appl. Meteorol., 28, 309-318.

Liou, K.N., 1980. An introduction to atmospheric radiation. Academic press. pp 183.

Mishchenko M. I. and L. D. Travis, 1997. Satellite retrieval of aerosols properties using polarization as well as intensity of reflected sunlight. J. Geophys. Res. 102, D14, 16989-17013.

Moulin, C., F. Guillard, F.Dulac and C.E.Lambert 1997. Long Term daily monitoring of Saharan dust load over ocean using METEOSAT ISCCP-2 data, I, methodology an dpreliminary results for 1983-1994 in the Mediterranean. J. Geophys. Res., 102, 16947-16969.

Nadal, F. and F.M. Breon, 1999. Parameterization of surface polarized reflectance derived from POLDEr spaceborne measurements. IEEE Trans. Geosc. Rem. Sens., 37, 1709-1718.

Nakajima, T. and A.Higurashi 1997: AVHRR Remote Sensing of aerosols optical properties in the Persian Gulf Region, summer 1991.J. Geophys. Res., 102, 16935-16946.

North, P.R.J., S.A.Briggs, S.E.Plummer and J.E.Settle, 1999. Retrieval of land surface bidirectional reflectance and aerosol opacity from ATSR-2 multiangular imagery. IEEE Trans. Geosc. and Remote Sens. 37, 526-537.

Rao, C.R.N., L.L.Stowe and E.P.McClain, 1989. Remote sensing of aerosols over ocean using AVHRR data: theory, practice and application. Int. J. Remote sens. 10, 743-749.

Roujean, J.L., M.Leroy and P.Y. Deschamps, 1992. A bidirectional reflectance model of the Earth surface for the correction of remote sensing data. J. Geophys. Res., 97, D18, 20455-20468.

Shenk, W.L., and R.J.Curran 1974. The detection of dust storms over land and water water with satellite visible and infrared measurements. Mon. Wea. Rev., 102, 830-837.

Shifrin, K.S., 1995. Simple relationships for the Angstrom parameter of disperse systems. Appl. Opt., 34, n.21, 4480-4485.

Siegel, D.A., M. Wang, S. Maritorena and W. Robinson, 2000. Atmospheric correction of satellite ocean color imagery: the black pixel assumption. Appl. Opt., 39, n.21, 3582-3591.

Tanre, D., Y.J.Kaufman, M.Herman and S.Mattoo, 1997. Remote sensing of aerosol properties over oceans using the MODIS/EOS spectral radiances. J. Geophys. Res. 102, D14, 16971-16988.

Tanre D., M. Herman, Y.J. Kaufman, 1996 Information on the Aerosol Size Distribution contained in the Solar Reflected Spectral Radiances, J. Geophys. Res., 101, 19043-19060.

Tanre, D. and M. Legrand, 1991. On the satellite retrieval of Saharan dust optical thickness over land: Two different approaches. J. Geophys. Res., 96, 5221-5227.

Tonna G., T. Nakajima and A. Roa: Aerosol features retrieved from solar aureole data: A simulation study concerning a turbid atmosphere, Appl. opt., 34, n 21, 4486-4499.

Torres, O., P.K.Barthia, J.R.Herman, Z.Ahmad and J.Gleason, 1998. Derivation of aerosol properties from satellite measurements of backscattered ultraviolet radiation: Theoretical basis. J. Geophys. Res., 103, 17099-17110.

Veefkind, J.P. and G. de Leeuw, 1998. A new aerosol retrieval algorithm to determine the spectral aerosol optical depth from satellite measurements. J. of Aerosol Sci., 29, 1237-1248.

Veefkind, J.P., G. de Leeuw, P.A. Durkee, P.B. Russell, P.V. Hobbs, and J.M. Livingston, 1999. Aerosol optical depth retrieval using ATSR-2 and AVHRR data during TARFOX. J. Geophys. Res., 104, 2253-2260.

Viollier, M., D.Tanre and P.Y.Dechamps, 1980. An algorithm for remote sensing of water color from space. Boundary Layer Meteor., 18, 247-267.

Wald, A.E., Kaufman,Y.J., D.Tanre, and B.-C. Gao., 1998. Daytime and nighttime detection of mineral dust over desert using infrared spectral contrast. J. Geophys. Res., 103, 32307-32313.

Chapter 3

Ocean Remote Sensing by Microwave and Visible-Infrared Sensors

Sea Modeling by Microwave Altimetry

STEFANO PIERINI
Dipartimento di Fisica (INFM), Università dell'Aquila, Italy.

Key words: Altimeter, geostrophic currents, ocean modelling

Abstract: In this note the applications of altimeter data to the monitoring of the sea surface height and, in particular, to the evaluation of large-scale oceanic currents are reviewed. The general problem of the altimeter measurement, with a special reference to the TOPEX/POSEIDON mission, is discussed. The decomposition of the sea surface height anomaly into its various components (tides, inverted barometer, steric height, height associated to geostrophic currents) is then considered. For each component a brief description of the physics of the associated phenomenon is given, along with references to the related recent scientific literature. A special emphasis is put on the determination of oceanic geostrophic currents. Therefore, the methods used to eliminate the components of the sea surface height anomaly not related to geostrophy are discussed. The derivation of near-surface currents and of depth-dependent currents with the aid of hydrographic data is considered in some detail. Finally, oceanic wind-driven models, in particular the time-dependent Sverdrup balance, are discussed in connection with altimeter data.

1. INTRODUCTION

The altimetry is a spaceborne microwave radar that is able to measure with an extremely high accuracy the distance between the satellite carrying the instrument and the sea surface. This has very important implications concerning different aspects of the earth sciences, such as the accurate determination of the earth shape (including the ocean bathymetry), the analysis of the oceanic tides and the monitoring of relevant aspects of the ocean circulation. In Section 2 the general problem of altimeter measurements, with particular attention to the TOPEX/POSEIDON mission

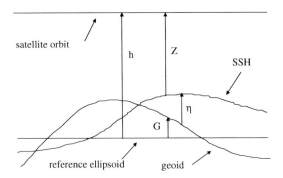

Figure 1. Definition of the parameters h, G, Z and η.

and to the current inadequacy of altimeter data to study mean currents, will be reviewed. In this note we will put the stress on oceanographic implications, namely on the determination from altimeter data of time-varying geostrophic currents associated to the dynamically active part of the sea-surface height (SSH) anomaly. From our point of view we will therefore consider the remaining part of the altimeter signal as a "noise" to be filtered out. Section 3 is devoted to the analysis of these effects and to the methods that can be adopted in order to remove them. In section 4 the computation of near-surface geostrophic currents from the dynamically active SSH anomaly, and of the current vertical shear with the aid of hydrographic data are discussed. In the same section the observation of Rossby waves with the altimeter is also briefly considered. Finally, in Section 5 oceanic wind-driven models, with particular emphasis on the time-dependent Sverdrup balance, are considered in connection with altimeter data.

2. ALTIMETER DATA AND THE VARIABILITY OF THE SSH

The general problem posed by the altimeter measurements can be summarized as follows (e.g., Robinson, 1994). Let us define a "reference ellipsoid" as the one that approximates the shape of the earth surface without considering small scale anisotropies in the distribution of mass. This ellipsoid has R_{min}=6357 km and R_{max}=6378 km, where R_{min} is the polar radius and R_{max} is the equatorial radius. The satellite orbit (h), and the "geoid" (G) are referred to this reference surface. The geoid is the geopotential surface that would coincide with the sea surface if the sea were in equilibrium, and presents small scale departures from the reference ellipsoid (ranging from -104 m to +64 m) because anomalies in the distribution of mass are present. It is therefore natural to refer the sea-surface height (η) to the geoid, and finally let us denote as Z the altimeter data measured by the radar (Fig. 1).

The oceanographically interesting quantity η, referred to a given point of the sea surface, will therefore be given by:

$$\eta(t) = h - G - Z(t) \ . \tag{1}$$

Let us now discuss briefly the errors associated to each quantity. Starting from the Skylab mission (1973-1974), passing through GEOS 3 (1975-1978), Seasat (1978), Geosat (1985-1989), ERS 1/2 (1991-2000/1995) and arriving to the joint United States/French mission TOPEX/POSEIDON (from now on denoted as T/P), the accuracy in the determination of each of these variables has greatly improved. For the T/P data, which are the ones we will be mainly concerned with in this note, both the satellite orbit determination h and the altimeter measurement Z are affected by an error of only 3-5 cm, which is astonishingly small if compared to the orbital altitude of O(1000 km). One should bear in mind that to achieve such a high degree of accuracy for h and Z it is not only necessary to have a state-of-the-art radar, but it is also fundamental to introduce appropriate corrections for atmospheric transmission and surface roughness errors and, as far as h is concerned, it is necessary to rely on a satisfactory satellite orbital model which, fortunately, does not depend heavily on the small scale variations of the geoid.

Before proceeding to the analysis of the errors, let us spend few words on the T/P mission. The satellite was launched in August 1992 and was expected to operate through September, 1998, although the satellite was operational for a longer time (see the special section of the *Journal of Geophysical Research*, 99 (C12), 24,369-25,062, 1994, devoted to "TOPEX/POSEIDON: Geophysical Evaluation" for general aspects of the mission, the gravity model, the orbit determination and their accuracy). The orbital altitude was chosen relatively high (1336 km) in order to reduce atmospheric drag and gravity forces acting on the satellite, and therefore maximizing the accuracy of orbit determination. The repeat period (9.916 days), the inclination (66 degrees), and the cross-track separation (316 km at the equator, 240 km at 40°) insure a good compromise for the temporal and spatial resolutions, and avoid undesirable aliasing problems for the tidal constituents (Fu et al., 1994).

Now, if the geoid G were affected by an error comparable or smaller than that of h and Z, then the altimeter mapping of a given sea region would provide the SSH within an error of 3-5 cm. In particular, as the repeat period of T/P is ~10 days, a large scale oceanic region would be covered in a time smaller than most oceanic variability, therefore the obtained η over that area could be seen as an "instantaneous" picture of the sea surface topography and (as we will see in sect. 4), indirectly, of geostrophic currents. However, the geoid is still affected by an error larger than 3-5 cm, ranging from more than 20 cm for relatively small spatial scales (the interesting ones for oceanographic applications) to 10 cm for scales larger than 2000 km, and this makes the determination of the instantaneous currents inaccurate. Let us consider, for instance, the JGM-3 Ohio State University (OSU91a) geoid model developed by the T/P project (Rapp et al., 1991; Nerem et al., 1994; Tapley et al., 1996). Stammer et al. (1996) computed the mean SSH averaged over a 2-year period from T/P data relative to this geoid model (in other words they took a long time average of (1), obtaining a given <η>). They then subtracted this "measured" mean SSH from the one ($<\eta>_{model}$) computed by the global Parallel Ocean Climate Model (POCM) of Semtner and Chervin (1992), which in this framework is considered as a reference "real" SSH. Major differences were indeed found primarily on relatively small scales, especially in oceanic trench systems where the geoid is known to be affected by large error estimates. It should be mentioned, however, that for particularly strong currents, such as for instance the Antarctic Circumpolar Current (ACC), the altimeter can provide information also on the mean flow (e.g., Park and Gambéroni, 1995). Recently a method to determine the mean SSH from altimeter data based on the knowledge of its temporal variability was proposed by Feron et al. (1998). It gives promising results for ocean areas with strong mesoscale variability, such as the western boundary currents and the ACC. However, in general, up to now geoid estimates appear to have proved inadequate to use altimetric data for improving the existing knowledge of the average large-scale oceanic circulation. A sufficiently accurate evaluation of the geoid for the computation of absolute currents is nevertheless expected in the near future.

On the other hand, in order to obtain interesting oceanographic information from Z one can take advantage of the fact that the satellite orbit is repeating, i.e. the ground tracks are repeatedly covered by the satellite, so that long time series are available. One can therefore define the SSH and altimeter measurement anomalies η' and Z', respectively, as follows:

$$\eta(t) = \langle \eta \rangle + \eta'(t), \quad Z(t) = \langle Z \rangle + Z'(t),$$

where <.> represents time averaging. Thus, by taking the average of (1) and subtracting the means one finally obtains:

$$\eta'(t) = -Z'(t). \tag{2}$$

This equation shows that the temporal anomaly of the SSH can be evaluated with the same accuracy of h and Z no matter how accurate is the geoid determination, the latter being eliminated by the averaging procedure along with the mean value of the SSH.

3. DECOMPOSITION OF THE SSH ANOMALY INTO ITS VARIOUS COMPONENTS

The dynamically active part (η'_D) of the SSH anomaly we are seeking for, i.e. the one related to geostrophic currents, is only one of the components into which the measured anomaly η' can be decomposed. One can write:

$$\eta' = \eta'_T + \eta'_P + \eta'_S + \eta'_D + \delta\eta', \qquad (3)$$

where η'_T is the SSH anomaly related to the ocean tides, η'_P is related to the isostatic response of the ocean surface to atmospheric pressure fluctuations, η'_S is the "steric" anomaly associated to the expansion or contraction of water due to heat exchange with the atmosphere and, finally, $\delta\eta'$ is the anomaly associated to errors and to all the other (mainly small scale) non-geostrophically balanced motions. Although (3) includes motions that have very different space and time scales (Wunsch and Stammer, 1995), nonlinear interactions can nonetheless be present. However, they are usually not very relevant, and here we assume that each term can be considered separately. In the remainder of this section the first three anomalies in (3) will be briefly considered. In the next two sections the dynamically active part of the signal will be discussed in more detail.

3.1 Tides

The knowledge of the ocean tides has greatly improved in the last two decades thanks both to the unprecedented accuracy and global coverage of the altimeter data, and to the implementation of numerical world tidal models. The tides are the only oceanic motions that can, in principle, be predicted provided the amplitudes and phases of all the consituents are known. The combination of altimeter data and models has, in fact, produced most accurate global maps of the tides which now have an accuracy of 2 cm in the deep ocean (Shum et al., 1997). The term η'_T can therefore be eliminated from (3), without modifying the accuracy of the balance, by means of state-of-the-art tidal models (e.g., Ma et al., 1994; Le Provost et al.,

1994; Eanes and Bettadpur, 1995; Andersen at al., 1995). A brief discussion about the observation of tides with the altimeter is presented below.

In points along a repeat ground track altimeter data provide time series of η' that can be analysed by means of a spectral analysis, in a manner analogous to what is usually done with tide gauge data. Since the tidal frequencies are known, if the time series are long enough to resolve all the frequencies, the amplitudes and phases of the constituents can be determined. For altimeter data problems arise because the observed frequencies are not the effective tidal ones but they are alias frequencies because the signal is sampled at intervals longer than half the periods (for T/P we have seen that the cycle repeats every ~10 days while the main tidal components are semidiurnal and diurnal). To give an example, the T/P alias periods for M_2 (12.42 h), S_2 (12.00 h), K_1 (23.93 h), and O_1 (25.82 h) are: 62.11, 58.74, 173.19, 45.71 days, respectively (Schlax and Chelton, 1994). The time series must therefore be particularly long in order to resolve such low frequencies, but the long time coverage of T/P is sufficient to resolve the main constituents. For this kind of data the tidal analysis known as the "response method" is usually applied (e.g., Munk and Cartwright, 1966; Cartwright and Ray, 1990; Andersen, 1995). For the analysis of the global ocean tides in the deep ocean with T/P one can refer to Desai and Wahr (1995), Sanchez and Pavlis (1995), Andersen (1995), Andersen et al. (1995), Kantha (1995), Kantha et al. (1995), Matsumoto et al. (1995), Arnault and Le Provost (1997), Desai et al. (1997), and Tierney et al. (1998). Finally, it should be mentioned that, since in shallow coastal seas the tides are largely dependent on the bathymetry and the shape of the coasts, the resulting horizontal length scales can be very small, so that the altimeter cannot resolve the tides properly. Nevertheless reliable empirical estimates can be obtained from T/P by combining along-track and crossover observations (Andersen, 1999). The use of coastal ocean models can also prove very useful (Foreman et al., 1998).

3.2   Inverted Barometer

Let us suppose that in the open ocean a stationary horizontal atmospheric pressure gradient is present at the sea surface due to a perturbation p_a superimposed on the spatially averaged atmospheric pressure. The ocean is in equilibrium (absence of motion) only if no horizontal pressure gradient in the sea is present. Therefore the sea level must adjust in such a way that the corresponding pressure anomaly balances the one associated to the atmospheric pressure (e.g., Gill, 1982), i.e.:

$$\rho g \eta_P(x, y) = -p_a(x, y), \qquad (4)$$

where ρ is the surface water density and g is the acceleration of gravity (for the sake of simplicity, rectangular coordinates are considered: x and y are horizontal coordinates on a tangent plane and z is the vertical coordinate). The SSH anomaly implied by (4) is that of an "inverted barometer", i.e. the local increase of the atmospheric pressure (e.g. of 1 mbar) is balanced by a depression of the sea surface (of 1 cm), so that at any depth in the sea no pressure difference is found (naturally, the sea surface topography is produced, during the transient, by the motion of water masses driven by the not yet balanced horizontal pressure gradient). Relation (4) is rigorously valid for a stationary pressure anomaly in an infinite ocean, but it applies also to realistic situations in which the pressure varies over periods longer than ~2 days (e.g., Ponte, 1992, 1993; Wunsch and Stammer, 1997). For these sufficiently long periods one can consider a quasi-isostatic balance given by (4) but where η and p are substituted by the fluctuations η' and p':

$$\rho g \eta'_p(x,y,t) = -p'_a(x,y,t). \quad (5)$$

The T/P altimeter data have proved a powerful tool to test (4-5) and to study its departures (Fu and Pihos, 1994; Gaspar and Ponte, 1997; Ponte and Gaspar, 1999). On the other hand, the knowledge of the surface atmospheric pressure, provided for instance by the NCEP or ECMWF meteorological analysis, allows to remove, through (5), the term η'_p in (3) (e.g. Wunsch and Stammer, 1997; Stammer, 1997). It should be noticed that in a semi-enclosed sea such as the Mediterranean basin the assumption that the ocean be infinite is far from being satisfied, therefore more complex models than (5) should be adopted (e.g., Candela, 1991; Le Traon and Gauzelin, 1997).

For atmospheric pressure variations over time scales smaller than ~2 days the ocean does not respond isostatically, i.e. through (5), and even for periods longer than 2 days, oceanic rotational normal modes can be excited (e.g., Ponte, 1993, 1997; Pierini, 1996; Pierini et al, 2002) leading to time-dependent SSH signals. Therefore, it should be borne in mind that part of the oceanic response to the atmospheric pressure variability in terms of SSH is included in the last term $\delta\eta'$ in (3).

3.3 Steric height

The main cause of the variability of the SSH outside the tropics is the density variation of the water column (without change of mass) associated to the net surface heat flux. This is the so-called "steric height" change, denoted by η'_s in (3), and given by (Gill and Niiler, 1973):

$$\eta'_s = -\frac{1}{\rho_0}\int_{-H}^{0}\rho' dz,\qquad(6)$$

where ρ' is the density anomaly, ρ_0 is a reference mean density and H is the water depth, although the main changes take place in the thin surface mixed layer which is affected by atmospheric buoyancy fluxes (there is also an adiabatic steric effect associated with the vertical displacement of isotherms due to the wind, but we do not consider this effect here). The SSH steric anomaly has a pronounced seasonal variability and manifests itself over large spatial scales of O(1000 km). It increases with increasing latitude, and the T/P data have shown a somewhat surprising asymmetry in amplitude between the two hemispheres (Stammer, 1997). This signal is dynamically passive, because below the mixed layer no horizontal pressure gradient is associated to it, since no change in mass is produced. On the other hand, near the surface the SSH variations do cause pressure variations, but they are over such large scales that the corresponding horizontal pressure gradient force, and therefore the associated geostrophic currents, are negligible. The steric SSH anomaly is important *per se* because it allows to monitor the ocean heat storage (e.g., White and Tai, 1995; Chambers et al., 1998). However, how to get rid of this large anomaly when studying geostrophic currents? The most sophisticated method is the one adopted by Stammer (1997) and Vivier et al. (1999) who modelled the steric term (6) by using ECMWF net heat fluxes. The steric term is computed by Polito and Cornillon (1997) by using the NODC climatology and by Gilson et al. (1998) using a climatology consisting of XBT and XCTD profiles. One can also take advantage of the large-scale nature of the signal in order to filter it out by means of appropriate averaging procedures, as done by Isoguchi et al. (1997) and Witter and Gordon (1999).

4. GEOSTROPHIC CURRENTS AND ROSSBY WAVES

In this section we discuss the term η'_D in (3) associated with the geostrophic balance and for which, therefore, information on oceanic currents can be obtained. In order to do this we begin by reviewing some basic notions of dynamical oceanography. An oceanic current is said to be in geostrophic balance if the Coriolis force and the horizontal pressure gradient force balance out exactly. In this case the current velocity can be written in the *f*-plane as follows (e.g., Pedlosky, 1987; Gill, 1982; Pond and Pickard, 1983):

$$u = -\frac{\partial \psi}{\partial y}, v = \frac{\partial \psi}{\partial x},\qquad(7)$$

where the streamfunction ψ is given by:

$$\psi(x,y,z) = \frac{1}{\rho f} p(x,y,z),$$

(*p* is the sea pressure, ρ(x,y,z) is the water density and *f* is the Coriolis parameter). Equation (7) implies that geostrophic currents flow parallel to the isobars, and it is rigorously valid only if the acceleration vanishes, i.e. for rectilinear and stationary currents. However, for flows that are sufficiently large-scale, both in space and time (more precisely, for small Rossby number flows; see Pedlosky, 1987), equation (7) provides an excellent approximation, so that primed variables can be substituted in (7) (from now on the time dependence is understood to satisfy this requirement).

If currents near the sea surface (z=0) are considered, then *p'* can be expressed, through the hydrostatic relation, in terms of the SSH anomaly:

$$\psi'_{surface}(x,y,t) = \frac{g}{f} \eta'_D(x,y,t),\qquad(8)$$

where the subscript *D* in η' indicates that we are now dealing with the dynamically active part of the SSH (see (3)). Equation (7) with (8) allows to derive the geostrophic surface current anomalies from the knowledge of the SSH anomaly as obtained from the altimeter, after subtracting other components not related to geostrophy, as discussed in sect. 3. This is the main scientific motivation on which the altimeter missions are based. Considering the error in the determination of η (sect. 2), large-scale altimeter-derived currents are, up to now, affected by an error of 3-5 cm/s (Strub et al., 1997). Examples of computation of upper ocean geostrophic currents from altimeter data (often performed by computing the along-track derivative of η'_D in order to avoid mapping errors, thus obtaining the cross-track velocity) are given by Kelly and Gille (1990), Yu et al. (1995), Larnicol et al. (1995), Snaith and Robinson (1996), Rapp et al. (1996), Strub et al. (1997), Iudicone et al. (1998), and Vivier and Provost (1999).

If surface geostrophic currents are interesting, even more important is the knowledge of the current as a function of depth or, at least, of depth-integrated currents, because they correspond to net heat oceanic transports that contribute substantially to the world climate. The altimeter cannot provide any direct information about this, but complementary in situ measurements can be used in conjunction with altimeter data in order to evaluate geostrophic currents at depth. We therefore pass to give a brief discussion on this topic. In a hypothetical homogeneous incompressible ocean a geostrophic current would be depth-independent, because no other source of horizontal pressure gradient but the one provided by the sea surface topography exists, therefore (7-8) would give the current at any depth. On the other hand, real oceans are stratified, a typical stratification consisting of a surface mixed layer followed by a region of sharp temperature variation (the thermocline), below which an almost constant-density water is present (but very large departures from this idealized picture can be present; see, for instance, Pickard and Emery, 1982). The presence of stratification is usually accompanied, for the mean flow, by a baroclinic compensation, which produces a geostrophic current that is small or even vanishes at great depth. In fact, the inclination of the isopycnic surfaces, produced by various causes, corresponds to a depth-dependent "relative" pressure field and, in turn, to a relative current field, which is the oceanic counterpart of the atmospheric "thermal wind" (e.g., Holton, 1979). This baroclinic current adds to the barotropic current given by (7-8), thus yielding a total depth-dependent geostrophic current, which reduces to (7-8) at the surface.

For the wind-driven circulation in large oceans, baroclinic compensation is the result of an adjustment process in which long baroclinic Rossby waves radiate from the eastern boundary (Anderson and Gill, 1975). At mid-latitudes the time needed to achieve this spin-up is of the order of a decade, therefore while the mean flow is compensated, the circulation induced by time-dependent winds is on the contrary mainly barotropic for periods less than O(1 year) (the barotropic spin-up is much faster, requiring only a few days to be effective). As a consequence, the current fluctuations measured by the altimeter for the same periods at mid-latitudes are expected to be representative not only of the subsurface dynamics but, to some degree, also of the entire water column. However, for lower latitudes the baroclinic spin-up time is sensibly smaller (the speed of Rossby waves increases rapidly toward the equator). Therefore near the tropics also the time-dependent ocean circulation yields a relevant baroclinic component. Apart from large oceans, in many other circumstances where coastal and topographic effects are important, relative currents can exist also for the time-dependent ocean circulation. Thus, the altimeter-derived SSH anomaly should, in general (but particularly at low latitudes), be complemented with in situ hydrographic measurements. Aspects related to this point will now be considered.

Hydrographic (i.e. temperature, salinity and pressure, and therefore density) measurements are performed by research vessels over large ocean areas and provide the vertical shear of the relative current through the thermal wind relationship, but no absolute current value. A classical problem of dynamical oceanography is the determination of the absolute current profile by prescribing a value of the current at some depth. This is often done by assuming a "level of no motion" in which the current vanishes (Pond and Pickard, 1983), although this "classical" method is affected by large uncertainties. Another more reliable method would be, in principle, that of using a "level of known motion" where geostrophic currents are known. This method is perfectly suitable for altimeter data because, as we have seen, the latter provide geostrophic surface currents, i.e. at z=0. The knowledege of the vertical derivative of the current given by the thermal wind (if the density structure is known) can therefore give the current also at z<0. Usually, the seasonal and longer term variability is analysed with this method. Examples of computation of geostrophic currents at depth from altimeter and hydrographic data are given by Han and Tang (1999), Buongiorno Nardelli et al. (1999), and Vivier and Provost (1999). An alternative way of computing currents below the surface from altimeter data is proposed by Williams and Pennington (1999) who, for a given background stratification in the North Atlantic, constrain a thermocline model by T/P altimetry, in so obtaining a relationship between SSH and the depth-integrated transport. Hydrographic data can also be used in an inverse way: for instance, in order to perform a quantitative test for an oceanic general circulation model, Stammer et al. (1996) use the SSH obtained by a global inversion of hydrographic and nutrient data along WOCE (World Ocean Circulation Experiment) sections and compare it with both model and altimeter-derived SSH.

Important dynamical features of the large scale oceanic variability that can be detected by the altimeter are Rossby waves (e.g., Pedlosky, 1987, 1996; Gill, 1982). They represent the messengers of energy in large scale (subinertial) atmosphere and ocean dynamics. In the oceans Rossby waves provide the dynamical mechanism for the adjustment to the large scale atmospheric forcing, and play a fundamental role in phenomena such as the westward intensification of boundary currents, the oceanic response to large scale fluctuating winds, boundary current radiation, etc.. Rossby waves are barotropic or baroclinic quasi-geostrophic current oscillations, and owe their existence to the conservation of potential vorticity of fluid columns in an ambient in which the planetary vorticity varies with latitude (the planetary β-effect), but a topographic counterpart also exists. Since Rossby waves are quasigeostrophic oscillations, (7-8) apply to them, therefore they can be revealed synoptically by the altimeter. This is a really revolutionary improvement in Rossby wave observation since, before, only local hydrographic, currentmeter and XBT measurements could give indication of their existence. Perhaps the most striking example of Rossby wave observations with T/P altimeter data is provided by Chelton and Schlax (1996), but several other analyses of the same nature have been carried out (e.g., Polito and Cornillon, 1997; Maltrud et al., 1998; White et al., 1998; Döös, 1999; Vivier et al., 1999; Witter and Gordon, 1999; Cipollini, 2000; Kobashi and Kawamura, 2001).

5. ALTIMETER DATA AND OCEANIC WIND-DRIVEN MODELS

A more advanced and promising application of altimeter-derived geostrophic currents is their analysis in the framework of oceanic models, in which currents are derived from the atmospheric forcing by solving appropriate dynamical equations. The comparison between *measured* and *modelled* currents allows to validate both the measuring technique and the model; moreover, discrepancies found from the comparison can reveal inadequacies both in the experimental and in the modelling approach, and can suggest possible improvements. The models can span from simple, classical wind-driven models such as the Sverdrup balance (see below) to extremely sophisticated state-of-the-art oceanic general circulation models (OGCMs) forced, for example, by ECMWF or NCEP surface fluxes. OGCMs allow for a detailed comparison with altimeter-derived currents (e.g., Chao and Fu, 1995; Stammer et al., 1996; Stammer, 1997; Fukumori et al., 1998), but it is sometimes difficult to identify physical mechanisms and interpret discrepancies, if all possible effects are taken into account. On the other hand, simplified models can sometimes allow for a better physical interpretation of the results, but they do not take into account effects that may interact with the ones considered. The use of both simple and sophisticated models appears to be the best way to proceed, as it is often done in recent studies.

Let us consider the Sverdrup balance as a prototype of simple but extremely interesting oceanic wind-driven model. The direct effect of the surface winds on the ocean is the transfer of horizontal momentum in the form of surface Ekman currents confined in a thin upper layer of frictional influence (the upper Ekman layer), in which the total transport is normal to the wind direction. The curl of the wind stress induces a divergence of the Ekman transport: the result is a displacement of the sea surface, with consequent generation of depth-independent geostrophic currents, that can be accompanied (as discussed in the preceding section) by depth-dependent relative currents. This is, in brief, the basic mechanism through which winds generate geostrophic currents in the ocean, and any theory aimed at explaining the general wind-driven circulation, and in particular the structure of the subtropical gyres (with a narrow and intense *western boundary current* and a broad and weak equatorward return flow in the oceanic interior) must be based on it. The model proposed by Sverdrup (1947) was the first to provide a relatively satisfactory explanation of the flow in the oceanic interior (i.e., away from the western boundary), and can be represented in terms of a stationary balance between the negative rate of change of planetary vorticity for the equatorward moving water columns and the negative vorticity input provided by the anticyclonic wind system, giving rise to the meridional Sverdrup mass transport (e.g., Pond and Pickard, 1983; Pedlosky, 1996):

$$M_y = \rho v H = \frac{curl_z \tau}{\beta}, \qquad (9)$$

where ρ is the water density, β is the derivative of f with respect to latitude, y is the meridional coordinate, v is the vertically averaged meridional component of the velocity (including Ekman and geostrophic currents), H is the water depth, and τ is the surface wind stress. It is important to notice that the local balance (9) does not depend on neither the density structure nor the vertical distribution of velocity.

The possibility of using (9) (with τ given by meteorological analysis) for a comparison with the meridional transport fluctuations computed from altimeter data by means of (7-8), relies on several assumptions, the main ones being: (a) the Ekman transport should be small compared to the geostrophic transport (or, alternatively, the Ekman transport should be subtracted from (9)); (b) topographic effects in the potential vorticity balance should be negligible (or, alternatively, they should be taken into account in the so-called *topographic Sverdrup balance*); (c) SSH fluctuations associated to baroclinic motions should be negligible; (d) the stationary relation (9) should also apply to time-dependent currents (time-dependent Sverdrup balance, TDSB).

Such a comparison has been carried out in recent studies (Fu and Davidson, 1995; Stammer, 1997; Isoguchi et al., 1997; Vivier et al., 1999). Interesting information are obtained for horizontally averaged transports: for instance, a significant indication of the TDSB in the high latitude North Pacific is obtained (Fu and Davidson, 1995; Stammer, 1997). At low latitudes things are quite different, as assumption (c) fails to hold. In a zonal belt centered at 13°N in the Pacific Ocean, Stammer (1997) found that the zonally averaged meridional Sverdrup transport anomaly, Q_{SVE}, and the corresponding anomaly obtained from T/P altimeter data under the assumption of barotropic flow, Q_{ALT}, are highly correlated signals with an important seasonal component. However, Q_{SVE} leads Q_{ALT} with a phase lag of about 3 months, moreover the Q_{ALT} rms is about an order of magnitude larger than the Q_{SVE} rms. Pierini (2002a) has proposed a dynamical interpretation of this result. A 2-layer primitive equation ocean model with free surface was implemented in an idealized Pacific ocean and was forced with a synthetic wind field (based on COADS and ECMWF wind data) representing a fairly realistic seasonal variability. The zonally integrated Sverdrup and altimeter-equivalent meridional transport anomalies obtained from the wind and the numerical free surface signal, respectively, yield phase shifts and relative amplitudes in good agreement with the above mentioned altimeter observations. A theoretical interpretation of these results is provided in terms of beta-refracted low latitude baroclinic Rossby waves generated at the eastern boundary. It should be noted, however, that such large difference between Q_{SVE} and Q_{ALT} does not imply that the TDSB does not hold in the tropics, but simply that the SSH anomaly cannot locally be used in (7-8) to compute vertically integrated transports, because assumption (c) does not hold. In fact, in the same study it was shown that the TDSB does hold in the framework of the numerical model, as Q_{SVE} models correctly the vertically integrated meridional transport anomaly computed from the numerical solution.

In general, even when agreement between Sverdrup and altimeter-derived meridional transport fluctuations is good for the seasonal signal, it must not be necessarily so for higher frequencies (corresponding to periods of up to few months); in other terms the failure of assumption (d) mentioned above should be considered. In fact, it is usually assumed that for periods above 1 month the Sverdrup balance is valid for time-dependent wind forcing (Willebrand et al., 1980), whereas it was recently shown (Pierini, 1997; 1998) that for periods 1<T<4 months the oceanic response is in terms of westward intensified forced barotropic Rossby waves. This implies large departures from the TDSB over extensive western and central regions of the oceans. In a systematic analysis of the validity of the TDSB, in connection with altimeter data, Pierini (2002b) has found confirmation of this behaviour. Experimental evidence of westward confined barotropic Rossby waves based on T/P observations is provided, for example, by Kobashi and Kawamura (2001). In the same study by Pierini (2002b), topographic effects (related to the failure of assumption (b)) and effects associated to nonlinearities and to the complexity of the spatial structure of the wind stress curl have been considered as well. Finally, it should be pointed out that a very specific and advanced problem such as the assimilation of altimeter data into OGCMs has not been considered in this note.

5.1 Acknowledgments:

This work was supported by the "*Agenzia Spaziale Italiana*".

6. REFERENCES

Andersen, O.B., 1995: Global ocean tides from ERS 1 and TOPEX/POSEIDON altimetry. *J. Geophys. Res.*, **100**, 25,249-25,259.

Andersen, O.B., 1999: Shallow water tides in the northwest European shelf region from TOPEX/POSEIDON altimetry. *J. Geophys. Res.*, **104**, 7729-7741.

Andersen, O.B., P.L. Woodworth, and R.A. Flather, 1995: Intercomparison of recent ocean tide models. *J. Geophys. Res.*, **100**, 25,261-25,282.

Anderson, D.L.T., and A.E. Gill, 1975: Spin-up of a stratified ocean, with application to upwelling. *Deep-Sea Res.*, **22**, 583-596.

Arnault, S., and C. Le Provost, 1997: Regional identification in the Tropical Atlantic Ocean of residual tide errors from an empirical orthogonal function analysis of TOPEX/POSEIDON altimetric data. *J. Geophys. Res.*, **102**, 21,011-21,036.

Buongiorno Nardelli, B., R. Santoleri, D. Iudicone, S. Zoffoli, and S. Marullo, 1999: Altimetric signal and three-dimensional structure of the sea in the Channel of Sicily. *J. Geophys. Res.*, **104**, 20,585-20,603.

Candela, J., 1991: The Gibraltar Strait and its role in the dynamics of the Mediterranean Sea. *Dyn. Atmos. Oceans*, **15**, 267-299.

Cartwright, D.E., and R.D. Ray, 1990: Oceanic tides from Geosat altimetry. *J. Geophys. Res.*, **95**, 3069-3090.

Chambers, D.P., B.D. Tapley, and R.H. Stewart, 1998: Measuring heat storage changes in the equatorial Pacific: a comparison between TOPEX altimetry and Tropical Atmosphere-Ocean buoys. *J. Geophys. Res.*, **103**, 18,591-18,597.

Chao, Y., and L.L. Fu, 1995: A comparison between the TOPEX/POSEIDON data and a global ocean general circulation model during 1992-1993. *J. Geophys. Res.*, **100**, 24,965-24,976.

Chelton, D.B., and M.G. Schlax, 1996: Global observations of oceanic Rossby waves. *Science*, **272**, 234-238.

Cipollini, P., D. Cromwell, G.D. Quartly, and P.G. Challenor, 2000: Remote sensing of oceanic extra-tropical Rossby waves. In "*Satellites, Oceanography and Society*", D. Alpern Editor, Elsevier Science, 99-123.

Desai, S.D., and J. M. Wahr, 1995: Empirical ocean tide models estimated from TOPEX/POSEIDON altimetry. *J. Geophys. Res.*, **100**, 25,205-25,228.

Desai, S.D., J.M. Wahr, and Y. Chao, 1997: Error analysis of empirical ocean tide models estimated from TOPEX/POSEIDON altimetry. *J. Geophys. Res.*, **102**, 25,157-25,172.

Döös, K., 1999: Influence of the Rossby waves on the seasonal cycle in the Tropical Atlantic. *J. Geophys. Res.*, **104**, 29,591-29,598.

Eanes, R.J., and S.V. Bettadpur, 1995: The CSR3.0 Global Ocean Tide Model. *Publ. CSR-TM-95-06*, Center for Space Res. , Univ. of Texas at Austin.

Feron, R.C.V., W.P.M. De Ruijter, and P.J. Van Leeuwen, 1998: A new method to determine the mean sea surface dynamic topography from satellite altimeter observations. *J. Geophys. Res.*, **103**, 1343-1362.

Foreman, M.G.G., W.R. Crawford, J.Y. Cherniawsky, J.F.R. Gower, L. Cuypers, and V.A. Ballantyne, 1998: Tidal correction of TOPEX/POSEIDON altimetry for seasonal sea surface elevation and current determination off the Pacific coast of Canada. *J. Geophys. Res.*, **103**, 27,979-27,998.

Fu, L.L., E.J. Christensen, C.A. Yamarone, Jr., M. Lefebvre, Y. Ménard, M. Dorrer, and P. Escudier, 1994: TOPEX/POSEIDON mission overview. *J. Geophys. Res.*, **99**, 24,369-24,381.

Fu, L.L., and R.A. Davidson, 1995: A note on the barotropic response of sea level to time-dependent wind forcing. *J. Geophys. Res.*, **100**, 24,955-24,963.

Fu, L.L., and G. Pihos, 1994: Determining the response of sea level to atmospheric pressure forcing using TOPEX/POSEIDON data. *J. Geophys. Res.*, **99**, 24,633-24,642.

Fukumori, I., Raghunath, R., and L.L. Fu, 1998: Nature of global large-scale sea level variability in relation to atmospheric forcing: A modeling study. *J. Geophys. Res.*, **103**, 5493-5512.

Gaspar, P., and R.M. Ponte, 1997: Relation between sea level and barometric pressure determined from altimeter data and model simulations. *J. Geophys. Res.*, **102**, 961-971.

Gill, A.E., 1982: *Atmosphere-Ocean Dynamics*. Academic Press, 662 pp.

Gill, A.E., and P.P. Niiler, 1973: The theory of the seasonal variability in the ocean. *Deep Sea Res.*, **20**, 141-177.

Gilson, J., D. Roemmich, B. Cornuelle, and L.L. Fu, 1998: Relationship of TOPEX/POSEIDON altimetric height to steric height and circulation in the North Pacific. *J. Geophys. Res.*, **103**, 27,947-27,965.

Han, G., and C.L. Tang, 1999: Velocity and transport of the Labrador Current determined from altimetric, hydrographic, and wind data. *J. Geophys. Res.*, **104**, 18,047-18,057.

Holton, 1979: *An Introduction to Dynamic Meteorology*. Academic Press.

Isoguchi, O., H. Kawamura, and T. Kono, 1997: A study on wind-driven circulation in the subarctic North Pacific using TOPEX/POSEIDON altimeter data. *J. Geophys. Res.*, **102**, 12,457-12,468.

Iudicone, D., R. Santoleri, S. Marullo, and P. Gerosa, 1998: Sea level variability and surface eddy statistics in the Mediterranean Sea from TOPEX/POSEIDON data. *J. Geophys. Res.*, **103**, 2995-3011.

Kantha, L.H., 1995: Barotropic tides in the global oceans from a nonlinear tidal model assimilating altimetric tides. 1. Model description and results. *J. Geophys. Res.*, **100**, 25,283-25,308

Kantha, L.H., C. Tierney, J.W. Lopez, S.D. Desai, M.E. Parke, and L. Drexler, 1995: Barotropic tides in the global oceans from a nonlinear tidal model assimilating altimetric tides. 2. Altimetric and geophysical implications. *J. Geophys. Res.*, **100**, 25,309-25,317.

Kelly, K.A., and S.T. Gille, 1990: Gulf Stream surface transport and statistics at 69°W from the Geosat altimeter. *J. Geophys. Res.*, **95**, 3149-3161.

Kobashi, F., and H. Kawamura, 2001: Variation of sea surface height at periods of 65-220 days in the subtropical gyre of the North Pacific. *J. Geophys. Res.*, **106**, 26,817-26,831.

Larnicol, G., P.Y. Le Traon, N. Ayoub, and P. De Mey, 1995: Mean sea level and surface circulation variability of the Mediterranean Sea from 2 years of TOPEX/POSEIDON altimetry. *J. Geophys. Res.*, **100**, 25,163-25,177.

Le Provost, C., M.L. Genco, F. Lyard, P. Vincent, and P. Canceil, 1994: Spectroscopy of the world ocean tides from a hydrodynamic finite element model. *J. Geophys. Res.*, **99**, 24,777-24,797.

Le Traon, P.Y., and P. Gauzelin, 1997: Response of the Mediterranean, mean sea level to atmospheric pressure forcing. *J. Geophys. Res.*, **102**, 973-984.

Ma, X.C., C.K. Schum, R.J. Eanes, and B.D. Tapley, 1994: Determination of ocean tides from the first year of TOPEX/POSEIDON altimeter measurements. *J. Geophys. Res.*, **99**, 24,809-24,820.

Maltrud, M.E., R.D. Smith, A.J. Semtner, and R.C. Malone, 1998: Global eddy-resolving ocean simulations driven by 1985-1995 atmospheric winds. *J. Geophys. Res.*, **103**, 30,825-30,853.

Matsumoto, K., M. Ooe, T. Sato, and J. Segawa, 1995: Ocean tide model obtained from TOPEX/POSEIDON altimetry data. *J. Geophys. Res.*, **100**, 25,319-25-330.

Munk, W.H., and D.E. Cartwright, 1966: Tidal spectroscopy and prediction. *Philos. Trans. R. Soc. London*, A, **259**, 533-583.

Nerem, R.S., et al., 1994: Gravity model development for TOPEX/Poseidon: Joint gravity model 1 and 2. *J. Geophys. Res.*, **99**, 24,405-24,448.

Park, Y.H., and L. Gambéroni, 1995: Large-scale circulation and its variability in the south Indian Ocean from TOPEX/POSEIDON altimetry. *J. Geophys. Res.*, **100**, 24,911-24,929.

Pedlosky, J., 1987: *Geophysical Fluid Dynamics*. Springer-Verlag, 710 pp.

Pedlosky, J. 1996: *Ocean Circulation Theory*. Springer-Verlag, 453 pp.

Pickard, G.L., and W.J. Emery, 1982: *Descriptive Physical Oceanography*. Pergamon Press, 249 pp.

Pierini, S., 1996: Topographic Rossby modes in the Strait of Sicily. *J. Geophys. Res.*, **101**, 6429-6440.

Pierini, S., 1997: Westward intensified and topographically modified planetary modes. *J. Phys. Oceanogr.*, **27**, 1459-1471.

Pierini, S., 1998: Wind-driven fluctuating western boundary currents. *J. Phys. Oceanogr.*, **28**, 2185-2198.

Pierini, S., 2002a: A model for the dynamical interpretation of the altimeter-derived seasonal variability in the North Pacific Ocean. Submitted.

Pierini, S., 2002b: An analysis of the validity of the time-dependent Sverdrup balance, with application to altimeter data. In preparation.

Pierini, S., A. Fincham, D. Renouard, R. D'Ambrosio, and H. Didelle, 2002: Laboratory modeling of topographic Rossby normal modes. *Dyn. Atmos. Oceans*, in press.

Polito, P.S., and P. Cornillon, 1997: Long baroclinic Rossby waves detected by TOPEX/POSEIDON. *J. Geophys. Res.*, **102**, 3215-3235.

Pond, S., and G.L. Pickard, 1983: *Introductory Dynamical Oceanography*. Pergamon Press, 329 pp.

Ponte, R.M., 1992: The sea level response of a stratified ocean to barometric pressure forcing. *J. Phys. Oceanogr.*, **22**, 109-113.

Ponte, R.M., 1993: Variability in a homogeneous global ocean forced by barometric pressure. *Dyn. Atmos. Oceans*, **18**, 209-234.

Ponte, R.M., 1997: Nonequilibrium response of the global ocean to the 5-day Rossby-Haurwitz wave in atmospheric surface pressure. *J. Phys. Oceanogr.*, **27**, 2158-2168.

Ponte, R.M., and P. Gaspar, 1999: Regional analysis of the inverted barometer effect over the global ocean using TOPEX/POSEIDON data and model results. *J. Geophys. Res.*, **104**, 15,587-15-601.

Rapp, R.H., Y.M. Ming, and N.K. Pavlis, 1991: The Ohio State 1991 geopotential and sea surface topography harmonic coefficient model. Rep. 410, Dep. of Geod. Sci. and Surve., Ohio State Univ..

Rapp, R.H., C. Zhang, and Y. Yi, 1996: Analysis of dynamic ocean topography using TOPEX data and orthonormal functions. *J. Geophys. Res.*, **101**, 22,583-22,598.

Robinson, I.S., 1994: *Satellite Oceanography*. Wiley, 455 pp.

Sanchez, B.V., and N.K. Pavlis, 1995: Estimation of main tidal constituents from TOPEX altimetry using a Proudman function expansion. *J. Geophys. Res.*, **100**, 25,229-25-248.

Schlax, M.G., and D.B. Chelton, 1994: Aliased tidal errors in TOPEX/POSEIDON sea surface height data. *J. Geophys. Res.*, **99**, 24,761-24,776.

Semtner, A.J., Jr., and R.M. Chervin, 1992: Ocean general circulation from a global eddy-resolving model. *J. Geophys. Res.*, **97**, 5493-5550.

Shum, C.K., et al., 1997: Accuracy assessment of recent ocean tide models. *J. Geophys. Res.*, **102**, 25,173-25,194.

Snaith, H.M., and I.S. Robinson, 1996: A study of currents south of Africa using Geosat satellite altimetry. *J. Geophys. Res.*, **101**, 18,141-18,154.

Stammer, D., 1997: Steric and wind-induced changes in TOPEX/POSEIDON large-scale sea surface topography observations. *J. Geophys. Res.*, **102**, 20,987-21,009.

Stammer, D., R. Tomakian, A. Semtner, and C. Wunsch, 1996: How well does a 1/4° global circulation model simulate large-scale oceanic observations? *J. Geophys. Res.*, **101**, 25,779-25,881.

Strub, P.T., T.K. Chereskin, P.P. Niiler, C. James, and M.D. Levine, 1997: Altimeter-derived variability of surface velocities in the California Current System. 1. Evaluation of TOPEX altimeter velocity resolution. *J. Geophys. Res.*, **102**, 12,727-12,748.

Sverdrup, H.U., 1947: Wind-driven currents in a baroclinic ocean, with application to the equatorial currents of the eastern Pacific. *Proc. Natl. Acad. Sci. USA*, **33**, 318-326.

Tapley, B.D., et al., 1996: The JGM-3 gravity model. *J. Geophys. Res.*, **101**, 28,029-28,049.

Tierney, C.C., M.E. Parke, and G.H. Born, 1998: An investigation of ocean tides derived from along-track altimetry. *J. Geophys. Res.*, **103**, 10,273-10,287.

Vivier, F., K.A. Kelly, and L.A.Thompson, 1999: Contributions of wind forcing, waves, and surface heating to sea surface height observations in the Pacific Ocean. *J. Geophys. Res.*, **104**, 20,767-20,788.

Vivier, F., and C. Provost, 1999: Volume transport of the Malvinas current: can the flow be monitored by TOPEX/POSEIDON? *J. Geophys. Res.*, **104**, 21,105-21,122.

White, W.B., Y. Chao, and C.K. Tai, 1998: Coupling of biennial oceanic Rossby waves with the overlying atmosphere in the Pacific basin. *J. Phys. Oceanogr.*, **28**, 1236-1251.

White, W.B., and C.K. Tai, 1995: Inferring interannual changes in global upper ocean heat storage from TOPEX altimetry. *J. Geophys. Res.*, **100**, 24,943-24,954.

Willebrand, J., S.G.H. Philander, and R.C. Pakanowski, 1980: The oceanic response to large-scale atmospheric disturbances. *J. Phys. Oceanogr.*, **10**, 411-429.

Williams, R.G., and M. Pennington, 1999: Combining altimetry with a thermocline model to examine the transport of the North Atlantic. *J. Geophys. Res.*, **104**, 18,269-18,280.

Witter, D.L., and A.L. Gordon, 1999: Interannual variability of South Atlantic circulation from 4 years of TOPEX/POSEIDON satellite altimeter observations. *J. Geophys. Res.*, **104**, 20,927-20,948.

Wunsch, C., and D. Stammer, 1995: The global frequency-wavenumber spectrum of oceanic variability estimated from TOPEX/POSEIDON altimetric measurements. *J. Geophys. Res.*, **100**, 24,895-24,910.

Wunsch, C., and D. Stammer, 1997: Atmospheric loading and the oceanic "inverted barometer" effect. *Rev. Geophys.*, **35**, 79-107.

Yu, Y., W.J. Emery, and R.R. Leben, 1995: Satellite altimeter derived geostrophic currents in the western tropical Pacific during 1992-1993 and their validation with drifting buoy trajectories. *J. Geophys. Res.*, **100**, 25,069-25,085.

Sea Scatterometry

MAURIZIO MIGLIACCIO
University of Naples "Parthenope", Naples – Italy.

Key words: microwave remote sensing, sea scattering, scatterometer, inverse problem

Abstract: This paper deals with the sea scatterometry, i.e. with the physical principles and techniques which permit to a scatterometer to perform radar measurements such as to allow the sea surface wind field retrieval. The paper aims at providing the necessary overview of the problem in question thus describing the fundamentals governing the physics of the approach and the relevant inversion technique

1. INTRODUCTION

Remote sensing is "…the science and art of obtaining useful information about an object, area or phenomenon through the analysis of data acquired by a device that is not in contact with the object, area, or phenomenon under investigation" (Lillesand and Kiefer, 1979) or more unconventionally it is "…legitimised voyeurism"!

Within the different forms of remote sensing the active microwave techniques have gathered more and more attention since they are able to illuminate the scene under selected conditions, e.g. frequency and polarization, and with a limited influence of the atmosphere (Elachi, 1987; Stewart, 1985; Ulaby et al., 1981). Apart from the engineering side of the problem the environmental remote sensing demands for remotely sensed data interpretation and this is not all an easy task due to the difficulty to relate the observed quantity with the geophysical quantity of interest.

In this paper we describe one of the few microwave remote sensing techniques which has shown the real capability to move from the scientific phase to the operational one although many research studies are still under development to improve the quality of the geophysical parameter estimation, e.g. (Bentamy et al., 1999; Liu et al., 1998; Quilfen et al., 2000). As a matter of fact, we hereafter describe the scatterometer i.e., a real aperture radar with great radiometric resolution vs. a limited spatial resolution (Elachi, 1987), and its application to retrieve sea surface wind field (Donelan and Pierson, 1987; Moore and Fung, 1979).

A scatterometer performs a set of (almost) simultaneous Normalized Radar Cross Section σ^o measurements at different azimuth angles and its most popular and effective application deals with the sea surface wind field retrieval (Elachi, 1987) also because the wind field knowledge is fundamental in many geophysical applications.

The paper is organized as follows. In Section II a brief historical perspective of the main civil satellite missions is provided. In Section III the physics underlying the scatterometer measurements is presented and in Section IV the inversion scheme which permits to determine the wind field from appropriately tailored scatterometer measuremts is detailed. The problem of the inversion scheme validation is also shown. Finally in Section V some conclusions are drawn.

2. AN HISTORICAL OVERVIEW

In this Section an essential overview of the main civil satellite scatterometer mission is provided. As a matter of fact the main goal of this Section is to outline the most important developments with special emphasis on what concern the applicative use of sea scatterometer data.

As a remarkable landmark of sea scatterometer we must first mention the SASS, i.e. the scatterometer mounted on board of the SEASAT satellite launched by the NASA in 1978 and that have been operating 100 days only (Allan, 1983a; Allan, 1983b; Barrick and Swift, 1980; Pierson, 1983). For the purposes of this Section we only note that the SASS is a Ku-band (14.6 GHz) fan-beam scatterometer composed by four antennas illuminating a symmetrical swath (475 km each) on both sides of the satellite ground track with a nadir gap of 400 km. As a consequence for each resolution cell two measurements are at disposal. Making reference to one side swath we have that the antennas pointing angles are 45° (fore beam) and 135° (aft beam). The measurement set-up allows a non-contemporary HH/VV polarization measurement. For further details the interested reader can refer to specialized literature, e.g. (Allan, 1983a; Grantham et al., 1977; Schroeder et al., 1982).

The follower of the SASS is the NSCAT a sea scatterometer mounted on board of the Japanese satellite ADEOS. The satellite has been launched the August, 16, 1996.

The NSCAT is still a Ku-band (14 GHz) fan-beam scatterometer but composed by six antennas illuminating a symmetrical swath (600 km each) on both sides of the satellite ground track with a nadir gap of 350 km (Naderi et al., 1991). As a consequence for each resolution cell three antennas perform independent measurements in this scheme. Making reference to one side swath we have that the antennas pointing angles are 45° (fore beam), 115° (mid beam) and 135° (aft beam). In this case the central antenna performs a true double-polarization measurement, i.e. it is able to contemporary measure the HH and VV normalized radar cross section. The NSCAT ended to operate in June 1997. More details on NSCAT can be found in (Naderi et al., 1991).

Not only the NASA has launched satellites carrying onboard sea scatterometers but also the European Space Agency (ESA). In particular, the WSC sensor has been mounted on board of the ERS-1/2 satellites, therefore the one on board of the ERS-2 satellite is currently in operation. The ERS-1 was launched in July 1991 while the ERS-2 was launched in April 1995. The WSC scatterometer is a C-band (5.3 GHz) fan-beam scatterometer composed by three antennas illuminating a one-sided swath (500 km) off-nadir of about 225 km (Attema, 1991). The antennas pointing angles are 45° (fore beam), 90° (mid beam) and 135° (aft beam). The WSC scatterometer is a VV-polarization sensor (Attema, 1991).

All previous sea scatterometers are fan-beam ones that is such that the antenna pointing angles are fixed with respect to the satellite track (Elachi, 1987). A novel class of sea scatterometer have been recently proposed and operated: the pencil-beam ones. In this case the sea surface multiple angular diversity measurements are perfomed by an antenna(s) electronic beam scanning, i.e. the antenna pointing angles are no more fixed during the satellite flight (Elachi, 1987).

Due to the premature ends of the operativity of the NSCAT the NASA has planned a quick recovery mission in order to fill the data gap. Such a mission is known as QuickSCAT and carries on board the SeaWinds sensor belonging to the class of pencil-beam scatterometers (Long and Spencer, 1997). The follower of such a mission is going to be the one operated on board of the ADEOS II satellite and carrying on a scatterometer twin of the SeaWinds one. The QuickSCAT satellite is an ad-hoc satellite launched the June, 19 1999 while the ADEOS II satellite is meant to be launched in 2001. The SeaWinds is a Ku-band (13.4 GHz) scatterometer. Two antenna beams are electronically scanned during the flight. The inner-beam is able to perform HH polarization measurements whereas the outer-beam performs VV polarization measurements (Long and Spencer, 1997). The inner-beam has a swath of about 1420 km and the outer one has a swath of about 1800 km. Both swaths have no nadir gap and therefore partially overlap. Due to the electronic beams scanning configuration the number of measurements performeed at various azimuth angles depends on the position of the resolution cell with respect to the across-track position (Oliphant and Long, 1999). As a matter of fact, due to the across-track resolution cell position a number ranging from one to four angular diversity measurements are at disposal (Oliphant and Long, 1999). As pointed in Section IV this directly affects the capability to determine the wind field direction. Much more details on this innovative scatterometer can be found in recent papers, e.g. (Long and Spencer, 1987; Spencer et al., 1997; Oliphant and Long, 1999).

The follower of the ESA WSC is going to be the ASCAT which will be mounted on board of the METOP three satellite constellation which will be started to be launched in 2003. It is meant to be a C-band (5.3 GHz) fan-beam scatterometer composed by six antennas illuminating a symmetrical swath (550 km each) on both sides of the satellite ground track with a nadir gap of 600 km. Making reference to the one-side swath the antennas pointing angles are 45° (fore beam), 90° (mid beam) and 135° (aft beam). Such a mission is meant to provide a denser spatial and temporal coverage of the sea and therefore is aimed at supporting not only scientific studies but operative applications. As a reference is useful to note that the WSC is able to furnish a global coverage after a 3 day period. Since all satellites are all polar orbiting sensor at approximately the same altitude the coverage of other sensors can be (approximately) determined only by comparing the swath size to the one pertaining to the WSC.

3. THE PHYSICS

In this Section we provide the fundamentals which physically explain the capability of the sea scatterometer to derive the wind field by appropriately tailored multiple angular diversity measurements.

Although studies regarding the sea scattering dates more than 50 years, they were mainly aimed at minimizing the sea clutter since undesired for military radar applications (Jakeman and Pusey, 1976). Use of radar for environmental monitoring is a much more modern topic of research. In a very essential summary, observations show that sea backscattering is characterized by:
a) a rapid response (seconds) to wind speed changes;
b) an increase with wind speed;
c) a decrease with incidence angle ϑ;
d) a sinusoidal-like behaviour with the azimuth angle φ, i.e. the angle formed by the wind and the surface projection of antenna beam ($\varphi = 0°$ when upwind).

In order to tackle the inversion problem, i.e the determination of the ocean surface wind field from an appropriate set of σ^o's, a direct model which relates the measured quantity σ^o with the geophysical quantity of interest, i.e the wind field, is needed. Before proceeding further we only note that even the definition of the wind field is not at all straightforward (Donelan and Pierson, 1987; Stewart, 1985; Stoffelen, 1998), in the following we consider it at 10 meters above the sea level.

Within such a framework it is useful the equivalence theorem (Franceschetti, 1997; Kong, 1990) which states that actual electromagnetic sources can be replaced by equivalent surface currents located over a geometrical bounding surface (Kong, 1990). If we consider a plane wave incident upon the random sea surface we get the following expression for the scattered field in the region of interest (Franceschetti, 1997; Kong, 1990; Valenzuela, 1978):

$$E_S(\mathbf{r}) = \frac{-jk\exp(-jkr)}{4\pi r}\left(\underline{\underline{I}}-\hat{k}_S\hat{k}_S\right)\cdot\iint_{S'}dS'\left\{\hat{k}_S\times\left[\hat{n}\times\mathbf{E}(\mathbf{r}')\right]+\zeta\left[\hat{n}\times\mathbf{H}(\mathbf{r}')\right]\exp(j\mathbf{k}_S\cdot\mathbf{r}')\right\} \quad (1)$$

wherein S' is the sea rough surface area on which the surface integration has to be carried out (Kong, 1990), $\hat{n}$ is the unit versor normal to S', ζ is the wave impedance in free space, $k=\omega\sqrt{\varepsilon_o\mu_o}$ is the electromagnetic wavenumber and k_s the corresponding scattering vector (Kong, 1990; Valenzuela, 1978). Different analytical approaches derive from eq.(1) once that proper approximations are conducted. For the purposes of sea scatterometer the most useful analytical approach is the small perturbation method (SPM) which assumes the surface variations are much smaller than the incident wavelength and the surface slopes are relatively small (Ulaby et. al. 1981; Chew and Fung, 1988).

Under the SPM approach we get to the following form of the normalized radar cross section in the backscattering case (Ulaby et. al. 1981):

$$\sigma_{pp}^0 = 16\pi k^4 \cos^4\vartheta|\alpha_{pp}|^2 W(\mathbf{k}=\mathbf{k}_B) \quad (2)$$

wherein ϑ the incident angle and α_{pp} is the pp-polarized effective Fresnel coefficient (Ulaby et al., 1981). $W(\cdot)$ is the directional sea spectrum (Guissard, 1993) and κ_B the Bragg wavevector (Guissard, 1993; Ulaby et al., 1981; Valenzuela, 1978). Obviously the sea spectrum is the Fourier transform of the displacement autocorrelation function $r_\eta(\cdot)$ (Guissard, 1993).

The directional sea spectrum $W(\kappa;\Phi)$ is decomposed in two parts: the omni-directional one $S(\kappa)$ and the spreading function $D(\kappa;\Phi)$, i.e.

$$w(k,\Phi) = \frac{1}{k} S(k) D(k,\Phi) \tag{3}$$

The scattering model which is obtained by combining eq.(2) and eq.(3) is able to show that the scattering is due to the centimeter (at microwaves) ocean waves, i.e. the ripples, and it is azimuth dependent due to the anisotropy of the sea spectrum. Therefore the SPM model is able to justify the observations and the need to perform simultaneous measurements at different pointing angles of the same resolution cell for permitting the determination of the wind field which is related to the sea spectrum (Donelan and Pierson, 1987; Moore and Fung, 1979, Ulaby et al., 1981). This model also shows the long wave tilting modulation (Moore and Fung, 1979, Ulaby et al., 1981), which is of major concern, and the long-wave short wave hydrodynamic modulation (Ulaby et al., 1981). As a consequence, a larger resolution cell, with respect to the Synthetic Aperture Radar (SAR), is here advisable to minimize such perturbations (Essen, 2000; Moore and Fung, 1979). A critical problem is the possible presence of a swell (Durden and Vesecky, 1985; Essen, 2000). For an anlytical scattering model aimed at the scatterometer inversion problem see also (Janssen et al., 1988).

Unfortunately, the scattering analytical models, although fundamentals to understand the physics governing the electromagnetic wave - sea surface interaction, are not currently able to predict a reliable quantitative relationship between the normalized radar cross section and the wind field because of some unsolved key physical modelling issues such as the general relationship between the wind field and the sea spectrum (Lemaire et al., 1999; Kerkmann, 1998; Moore and Fung, 1979; Ulaby et al., 1981).

In order to deal with these problems a set of semi-empirical geophysical model function (GMF) have been constructed. The GMF is a semi-empirical relationship between the observed quantity, $\sigma°$, with the geophysical quantity to be estimated, the wind field. As a matter of fact, the GMF, for a given scatterometer system is (Moore and Fung, 1979; Ulaby et al., 1981):

$$M = M(U, \varphi, f, p, \vartheta, un \bmod elled) = M(U, \varphi) \tag{4}$$

The GMF emphasizes the dependence of the normalized radar cross section with respect to the wind field, considers the radar parameter dependence such as the frequency and polarization while neglects other (hopefully) minor dependences (Donelan and Pierson, 1987; Freilich and Dunbar, 1993; Moore and Fung, 1979; Ulaby et al., 1981).

The GMF used by ESA are the CMOD series (now CMOD4) (Long, 1992; Stoffelen, 1998; Stoffelen and Anderson, 1997) while the GMF used by NASA are the relevant to the Ku-band (Wentz et al., 1984; Wentz and Smith, 1999), but other functional forms have been also proposed (Rufenach, 1995; Rufenach, 1998a; Rufenach et al., 1998).

Once a functional form has been suggested by preliminary physical studies the exact knowledge of the GMF calls for a calibration step. Such a step consists in the GMF free parameters determination by a best fitting with external (to the scatterometer) high quality colocated measurements (Long, 1992; Freilich and Dunbar, 1993; Rufenach, 1995; Stoffelen, 1998, Stoffelen and Anderson, 1997). Since the knowledge of an accurate GMF is fundamental for the estimation of accurate wind fields from the scatterometer measurements the calibration step must be carefully tackled (Freilich and Dunbar, 1993; Stoffelen, 1998). For example the spatial and temporal data colocation imposes tight restriction on the selection of radar and for example buoy measurements.

In order to get other details the reader is referred to specialized literature such as (Freilich and Dunbar, 1993; Long, 1990; Long, 1992; Migliaccio and Colandrea, 1998; Migliaccio et al., 1997; Rufenach, 1995; Stoffelen, 1998, Stoffelen and Anderson, 1997; Wentz et al., 1984; Wentz and Smith, 1999).

4. THE INVERSION SCHEME

In this Section we illustrate the problem of retrieving the wind field from multiple appropriately tailored measurements. The first part of the Section provide a simple approach to this problem in order to present the problem in question. This approach is similar to the one presented in (Elachi, 1987) and it obviously only useful for providing a first sketch of the problems to be dealt with. The second part of this Section illustrate a real point-wise inversion scheme. As a matter of fact, the first part of this Section is meant to be propedeutic to best understand real cases as the one detailed hereafter. In other words we move from an educational viewpoint of the problem to a realistic one.

As it has been illustrated in the former Section with the present state-of-art it is not possible to quantitively use the analytical scattering models and proper forms of semi-empirical model must be considered. Let us first assume the following form of the GMF, i.e of eq.(4),

$$M = AU^{\gamma}(1 + a\cos\varphi + b\cos 2\varphi) \tag{5}$$

wherein U is the wind speed and φ the azimuth angle, i.e. the angle between the antenna pointing direction and the wind field, whereas the other parameters are function of the observing system, i.e. frequency, polarization, incidence angle, etc (Donelan and Pierson, 1987; Stewart, 1985).

For sake of simplicity let us neglect the cos φ term which is relatively small (Elachi, 1987), i.e.

$$M \rightarrow M(U,\varphi) = AU^{\gamma}(1 + b\cos 2\varphi) \tag{6}$$

If we consider a SEASAT-like scatterometer we have at disposal for each resolution cell a couple of indipendent measurements separated by 90° and if we assume they follow the GMF we have:

$$\sigma_1^0 = AU^{\gamma}(1 + b\cos 2\phi) \tag{7}$$

and

$$\sigma_2^0 = AU^{\gamma}(1 + b\cos(2\phi + \pi)) = AU^{\gamma}(1 - b\cos 2\phi) \tag{8}$$

Hence combining σ_1^0 and σ_2^0 we have:

$$\sigma_1^0 \sigma_2^0 = 2AU^{\gamma} \tag{9}$$

and

$$\sigma_1^0 \sigma_2^0 = 2AU^{\gamma} b\cos 2\phi \tag{10}$$

Hence

$$(\sigma_1^0 - \sigma_2^0)/(\sigma_1^0 + \sigma_2^0) = b\cos 2\phi \tag{11}$$

As a consequence the two normalized radar cross section measurements allow to determine the wind speed, see eq.(9), and the wind direction with some ambiguity, see eq.(11). Namely ϕ, $-\phi$, $\phi+90°$ and $-\phi-90°$ are all possible solutions.

If we accomplish a third normalized radar cross section measurement 45° apart, say σ_3^0, as in the ERS-2 case, we have:

$$\sigma_3^0 = AU^\gamma(1+b\cos(2\phi+\pi/2)) = AU^\gamma(1+b\sin 2\phi) \qquad (12)$$

and the four-fold ambiguity can be reduced to a 180° one!

It is important to note that in this case, for sake of simplicity no noise has been considered and therefore it may appear that an additional measurement can shed light into the remaining ambiguity. This is actually untrue because of the GMF functional form and the presence of noise. Actually, such a 180° ambiguity is the best case that can be encountered in the inversion scheme (Kerkmann, 1998).

Such a simple example shows some important key issues that must be tackled into the inversion scheme procedure. As a matter of fact, any inversion scheme is made by two steps:
a) solution search (inversion step);
b) solution sorting (ambiguity removal step).

We move now to consider a real point-wise inversion scheme. A frame of $Nx \times Ny$ resolution cells is considered and for each one a function to be minimized is built up, i.e.

$$f_{ij}(U,\varphi) = \sum_{m=1}^{M} \left| \sigma_{ijm}^0 - M(U_{ij},\alpha_m+\varphi_{ij}) \right|^2 \qquad (13)$$

wherein σ_{ijm}^0 are the measurements performed at the ij-cell and that sum up to M. M is the considered GMF, U_{ij} and φ_{ij} are the wind speed and the wind direction (referred to a cartographic reference system) to be determined, α_m is the antenna pointing angle in the same reference system.

Actually eq.(13) can be properly adjusted whenever a priori information about the reliability of the single measurements is at disposal (Bartoloni et al., 1993; Oliphant and Long, 1999; Stoffelen, 1998).

The mathematical problem to be considered, for each resolution cell, is:

$$\min fij(U,\varphi) \qquad (14)$$

Due to the non-linearity of the problem and the presence of noise (measurement noise and model noise) the problem in question is not easy at all and can be tackled only by exploiting some physical properties of the problem. In order to cope to the difficulties of the problem we make some assumptions. In particular, we note that the GMF best works for higher wind speeds and that most likely a uniform wind field is present within the considered frame.

For the solution of our problem we employ a smart use of a local minimization algorithm (LMA) which is a computer time efficient routine but whose result depends on the starting point and a global one (GMA) which is not-efficient but able to find the global minimum with no need of starting point. For a wider discussion on this issue see (Migliaccio and Sarti, 1999).

The inversion scheme is briefly sketched as follows:
1. On the whole frame the LMA is applied with starting point (0,0);
2. The most reliable cell is selected by considering the cell in which at the step 1 the highest wind speed has been determined;
3. On the most reliable cell the GMA is applied. The resulting wind field, say U^*, φ^* is considered as the true average wind over the frame;
4. On the whole frame the LMA is applied with starting point (U^*, φ^*). As a result we have the RANK1 solution;
5. On the whole frame the LMA is applied with starting point (U^*, $\varphi^*+180°$). As a result we have the RANK2 solution;
6. a The sorting between different solutions is accomplished by analyzing the minima value (Bartoloni et al., 1993).
6 b The sorting is accomplished by means of rough meteorological information (Elachi, 1987).

An applicative example of this inversion scheme over ERS-1 WSC data is shown in Figs.1 and 2. The data pertains to the Black Sea. In Fig.1 and Fig.2 the RANK 1 solution and the RANK 2 solution are shown, respectively.

Note that obviously the RANK 1 solution is not simply a 180° offset of the RANK 2 solution. In order to sort the proper solution it is here possible to make reference to ECMWF analysis data shown in Fig.3.

This shows that the RANK 2 is the correct one and that the average ECMWF wind field does not possess the fine structure of an instantaneous wind field (Liu et al., 1998) and this is important for applications.

Before proceeding further we need to note that the physical assumptions at the basis of this approach can be modified according to the kind of meteorological phenomena we are looking for. Further, some recent modification of this approach which best deal with light wind regimes have been recently proposed (Migliaccio and Sarti, 1999) as well as modifications which permit to span among more solutions than two (Migliaccio and Sarti, 2000) in accordance with (Kerkmann, 1998).

The interested reader can find an interesting review of different wind field retrieval procedures from sea scatterometer data in (Chi and Li, 1988; Rufenach, 1998b) and some details about the one used for NSCAT by

Sea scatterometry 193

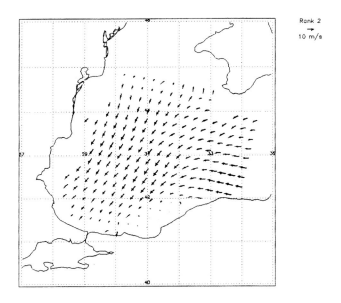

Fig.1: Relevant to the scatterometer WSC ERS-1 derived wind field over the Black Sea. RANK 1 solution.

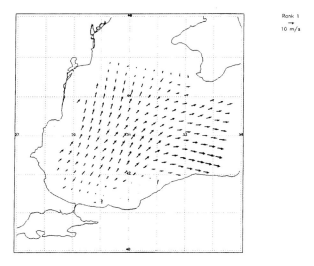

Fig.2: Relevant to the scatterometer WSC ERS-1 derived wind field over the Black Sea. RANK 2 solution.

NASA in (Shaffer et al.,1991). Within such a framework a novel approach based on the use of neural network is presented in (Chen et al., 1999) while another class of retrieval procedures are possible and they are based on the physical oceanographic model, e.g. the Navier-Stokes equations. This kind of approach is denoted as field-wise approach and although interesting has never been popular in operational terms. The reader can learn more on this by benefiting of the relevant paper of Long, i.e. (Long and Mendel, 1990).

In order to validate an inversion scheme a set of experiments can be carried on the basis of simulated, i.e. perfectly controlled, scenarios. With these respect we first make reference to the real point-wise procedure formerly illustrated and then we sketch some other interesting results excerpted from literature.

The first case we consider regards a set of simulations perfomed making use of the CMOD4 GMF (Long, 1992; Stoffelen, 1998; Stoffelen and Anderson, 1997), the ERS-1/2 configuration (Attema, 1991) and a proper level of noise (10%) (Bartoloni et al., 1993). The simulated wind field is characterized by different scale of complexity (Migliaccio et al., 2000). Such a complexity is measured by the uniformity ratio UR defined as the ratio between the wind field intensity mean $<U>$ and the root mean square intensity ΔU.

In Table I a set of various simulations are presented. In fact, some wind fields have been simulated with various UR and therefore the corresponding three σ^o have been generated according to a multiplicative noise model (Bartoloni et al., 1986; Bartoloni et al., 2000; Oliphant and Long, 1999). In particular, UR, $<U>$ and ΔU are listed together with the computer elapsing time and the accuracy of the reconstructed wind field measured according the ESA standards (± 2 m/s and $\pm 20°$).

Table I: Relevant to the accuracy of the inversion scheme. Simulation cases generated making use of the CMOD4 GMF and the ERS-1/2 configuration (Bartoloni et al., 2000; Migliaccio et al., 2000).

test	grid	$<U>$ [m/s]	ΔU [m/s]	UR	Time [s]	Acc. [%]
test1	20x19	7.2	4.5	1.6	31	94.47
test2	20x19	5.2	5.9	0.8	34	88.95
test3	15x19	10.7	4.9	2.2	21	95.44
test4	20x19	17.2	3.4	5.1	30	96.31
test5	20x19	13.8	2.0	1.6	24	96.05

This Table is only aimed at providing an idea of the accuracy that can be achieved by such an inversion scheme. The reader is referred to (Bartoloni et al., 2000) and (Migliaccio et al., 2000) for a wider discussion on such results.

Before ending this matter we like to underline that other kind of validations are possible although they are not at all straightforward and simple. In fact, it is conceptually easy to think of a comparison of wind field retrieved by means of scatterometer measurements with in situ data, e.g. buoys or vessel data. Although the reader is invited to refer to related litera-

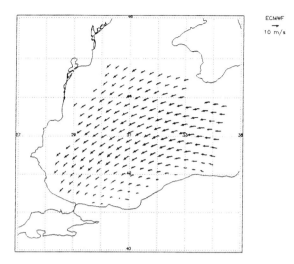

Fig.3: Relevant to the ECMWF analysis data.

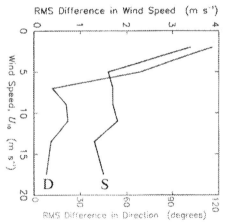

Fig.4: Relevant to a comparison of NSCAT reconstructed wind fields and vessel data (Bourassa et al., 1997). The line marked with S is relevant to the wind speed comparison while the line marked with D is relevant to the wind direction comparison.

ture for a wider discussion on the difficulties of such a validation, e.g. (Bourassa et al., 1997; Oliphant and Long, 1999; Marsili et al., 2000; Rufenach, 1998b), we like to present some results excerpted from (Bourassa et al., 1997) and obtained by analyzing NSCAT wind field and measurements performed on board of a research vessel, see Fig.4.

The interested reader can find other interesting validation studies in (Bentamy et al., 1999; Freilich and Dunbar, 1999; Kent et al., 1998; Marsili et al., 2000; Rufenach, 1998b)

5. CONCLUSIONS

In this paper an excursus on the physics rationale and inversion issues regarding the retrieval of sea surface wind field from scatterometer measurements has been reported and discussed. A large set of reference have been listed in order to ease further examination of the various aspects regarding sea scatterometry.

6. ACKNOWLEDGMENTS

The author wishes to thank the Italian Space Agency who partially supported this scientific project.

7. REFERENCES

Allan T.D. Ed., Satellite Microwave Remote Sensing, 1983a.
Allan T.D., "A Review of Seasat", in Satellite Microwave RemoteSensing, T.D.Allan Ed., pp.15-44, 1983b.
Attema, E.P.W., "The Active Microwave Instrument on-board the ERS-1 Satellite", Proc. IEEE, 79, pp.791-799, 1991.
Barrick, D.E. and C.T.Swift, "The Seasat microwave Instruments in Historical Perspective", IEEE J.Oceanic Eng., OE-5, pp.74-79, 1980.
Bartoloni, A., C.D'Amelio, and F.Zirilli, "Wind Reconstruction from Ku-band or C-band Scatterometer Data", IEEE Trans. Geosci. Remote Sensing, GE-31, pp.1000-1008, 1993.
Bartoloni, A., P.Colandrea, C.D'Amelio, M.Migliaccio, M.Sarti, "La Ricostruzione del Campo di Vento da Misure Scatterometriche", (in Italian) Istituto Universitario Navale - Volume di Studi in Memoria di A.Sposito, pp.243-254, 2000.
Bentamy, A., P.Queffeulou, Y.Quilfen and K.Katsaros, "Ocean Surface Wind Fields Estimated from Satellite Active and Passive Microwave Instruments", IEEE Trans. Geosci. Remote Sensing, GE-37, pp.2469-2486, 1999.
Bourassa, M.A., M.H.Freilich, D.M.Legler, W.Timothy Liu, and J.O'Brien, "Wind Observation from New Satellite and Research Vessels Agree", EOS, 78, pp.597-602, 1997.
Chen, K.S., Y.C.Tzeng, and P.C.Chen, "Retrieval of Ocean Winds from Satellite Scatterometer by a Neural Network", IEEE Trans. Geosci. Remote Sensing, GE-37, pp.247-256, 1999.
Chi, C.Y., and F.K.Li, "A Comparative Study of Several Wind estimation Algorithms for Spaceborne Scatterometers", IEEE Trans. Geosci. Remote Sensing, GE-26, pp.115-121, 1988.
Chew, M.F., and A.K.Fung, "A Numerical Study of the Regions of Validity of the Kirchhoff and Small Perturbation Rough Surface Scattering Models", Radio Sci., 23, pp.163-170, 1988.
Donelan, M.A., and W.J.Pierson, "Radar Scattering and Equilibrium Ranges in Wind-generated Waves with Application to Scatterometry", J.Geophys. Res., 92, pp.4971-5029, 1987.

Durden, S.P., and J.F.Vesecky, "A Physical Radar Cross Section Model for a Wind-Driven Sea with Swell", IEEE J.Oceanic Eng., OE-10, pp.445-451, 1985.

Elachi, C., Spaceborne Radar Remote Sensing: Applications and Techniques, IEEE Press, 1987.

Essen, H.H., "Theoretical Investigation on the Impact of Long Surface Waves on Empirical ERS-1/2 Scatterometer Models", Int. J.Remote Sensing, 21, pp.1633-1656, 2000.

Franceschetti, G., Electromagnetics, Plenum Press, 1997.

Freilich, M.H., and R.S.Dunbar, "Derivation Satellite Wind Model Functions Using Operational Surface Wind Analyses: An Altimeter Example", J.Geophys. Res., 98, pp.14633-14649, 1993.

Freilich, M.H., and R.S.Dunbar, "The Accuracy of the NSCAT 1 Vector Winds: Comparison with National Data Buoy Center buoys", J.Geophys. Res., 104, pp.11231-11246, 1999.

Grantham, W.L., E.M.Bracalente, W.L.Jones, and J.W.Johnson, "The Seasat-A Satellite Scatterometer", IEEE J.Oceanic Eng., OE-2, pp.200-206, 1977.

Guissard, A., "Directional Spectrum of the Sea Surface and Wind Scatterometry", Int. J. Remote Sensing, 14, pp.1615-1633, 1993.

Jakeman, E., and P.N.Pusey, "A Model for Non-Rayleigh Sea Echo", IEEE Trans. Antennas Propagat., AP-24, pp.806-814, 1976.

Janssen, P.A.E.M., H.Wallbrink, C.J.Calkoen, D. van Halsema, W.A.Oost, and P.Snoeij, "VIERS-1 Scatterometer Model", J.Geophys. Res., 103, pp.7807-7831, 1988.

Kent, E.C., P.K.Taylor, and P.G.Challenor, "A Comparison of Ship and Scatterometer Derived Wind Speed Data in Open Ocean and Coastal Areas", Int. J. Remote Sensing, 19, pp.3361-3381, 1998.

Kerkmann, J., Review on Scatterometer Winds, Eumetsat Tech. Rep. No.3, 1998.

Kong, J.A., Electromagnetic Wave Theory, J.Wiley, 1990.

Lemaire, D., P.Sobieski, and A.Guissard, "Full-Range Sea Surface Spectrum in Nonfully Developed State for Scattering Calculations", IEEE Trans. Geosci. Remote Sensing, GE-37, pp.1038-1051, 1999.

Lillesand, T.M. and R. W. Kiefer, Remote Sensing and Image Interpretation, John Wiley, 1979.

Liu, W.T., W.Tang, and P.S.Polito, "NASA Scatterometer Provides Global Ocean Surface Wind Fields with More Structures than Numerical Weather Prediction", Geophys. Res. Lett., 25, pp.761-764, 1998.

Long, A.E., "C-band V-polarized Radar Sea-Echo Model from ERS-1 Haltenbanken Campaign", Proc. URSI 1992, pp.113-118, 1992.

Long, D.G., and J.M.Mendel, "Model-Based Estimation of Wind Fields Over the Ocean From Wind Scatterometer Measurements, Part I: Development of the Wind Field Model", IEEE Trans. Geosci. Remote Sensing, GE-28, pp.349-360, 1990.

Long, D.G., and M.W.Spencer, "Radar Backscatter Measurement Accuracy for a Spaceborne Pencil-Beam Wind Scatterometer with Transmit Modulation", IEEE Trans. Geosci. Remote Sensing, GE-35, pp.102-114, 1997.

Marsili, S., M.Migliaccio, M.Sarti, "Cersat WNF data and Urania Campaigns", Proc. Ocean Winds, Brest (France), pp.64, 2000.

Migliaccio, M., and P.Colandrea, "On the Radar Cross Section Data Binning", IGARSS'98, Seattle, WA, pp.974-976, 1998.

Migliaccio, M., and M.Sarti, "A Study on the Inversion of Sea Scatterometer Measurements", Alta Frequenza , 4, pp.66-70, 1999.

Migliaccio, M., and M.Sarti, IUN Internal Report on Wind Field Reconstruction, 2000.

Migliaccio, M., M.Sarti, P.Colandrea, and A.Bartoloni, "Accuracy of a PC-based Fan-Beam Wind Scatterometer Measurements Inversion Scheme", IGARSS'00, Honolulu, HI (USA), pp.263-265, july 2000.

Migliaccio, M., P.Colandrea, A.Bartoloni, C.D'Amelio, "On a New Relationship Between Radar Cross Section and Wind Speed Using ERS-1 Scatterometer Data and ECMWF Analysis Data over the Mediterranean Sea", IGARSS'97, Singapore, pp.1853-1855, aug.1997

Moore, R.K., and A.K.Fung, "Radar Determinations of Winds at Sea", Proc. IEEE, 67, pp.1504-1521, 1979.

Naderi, F., M.H.Freilich, D.G.Long, "Spaceborne Radar Measurement of Wind Velocity over the Ocean - An Overview of the NSCAT Scatterometer System", Proc. IEEE, 79, pp.850-866, 1991.

Oliphant, T.E., and D.G.Long, "Accuracy of Scatterometer-Derived Winds using the Cramer-Rao Bound", IEEE Trans. Geosci. Remote Sensing, GE-37, pp.2642-2652, 1999.

Pierson, W.J., "Highlights of the SEASAT-SASS Program: A Review", in Satellite Microwave RemoteSensing, T.D.Allan Ed., pp.69-86, 1983.

Quilfen, Y., A.Bentamy, P.Delecluse, K.Katsaros, and N.Grima, "Prediction of Sea Level Anomalies Using Ocean Circulation Model Forced by Scatterometer Wind and Validation Using TOPEX/Poseidon Data", IEEE Trans. Geosci. Remote Sensing, GE-38, pp.1871-1884, 2000.

Rufenach, C.L., "A New Relationship between Radar Cross-section and Ocean Surface Wind Speed Using ERS-1 Scatterometer and Buoy Measurements", Int. J.Remote Sensing, 16, pp.3629-3647, 1995.

Rufenach, C.L., "ERS-1 Scatterometer Measurements - Part II: An Algorithm for Ocean-Surface Wind Retrieval Including Light Winds", IEEE Trans. Geosci. Remote Sensing, GE-36, pp.623-635, 1998a.

Rufenach, C.L., "Comparison of Four ERS-1 Scatterometer Wind Retrieval Algorithms with Buoy Measurements", J.Atmos. Oceanic Technol., 15, pp.304-313, 1998b.

Rufenach, C.L., J.J.Bates, and S.Tosini, "ERS-1 Scatterometer Measurements - Part I: The Relationship Between Radar Cross Section and Buoy Wind in Two Oceanic Regions", IEEE Trans. Geosci. Remote Sensing, GE-36, pp.603-622, 1998.

Schroeder, L.C., W.L.Grantham, J.L.Mitchell, and J.L. Sweet, "SASS Measurements of the Ku-band Radar Signature of the Ocean", IEEE J.Oceanic Eng., OE-7, pp.3-14, 1982.

Shaffer, S.J., R.S.Dunbar, S.V.Hsiao and D.G.Long, "A Median-Filter-Based Ambiguity Removal Algorithm for NSCAT", IEEE Trans. Geosci. Remote Sensing, GE-29, pp.167-173, 1991.

Spencer, M.W., C.Wu, and D.G.Long, "Tradeoffs in the Design of a Spaceborne Scanning Pencil Beam Scatterometer: Application to SeaWinds", IEEE Trans. Geosci. Remote Sensing, GE-35, pp.115-126, 1997.

Stewart, R.H., Methods of Satellite Oceanography, University of California Press, 1985.

Stoffelen, A., Scatterometry, PhD Thesis, 1998.

Stoffelen, A., and D.Anderson, "Scatterometer Data Interpretation: Estimation and Validation of the Transfer Function CMOD4", J.Geophys. Res., C102, pp.5767-5780, 1997.

Ulaby, F.T., R.K.Moore, and A.K.Fung, Microwave Remote Sensing, Addison-Wesley, 1981.

Valenzuela, G.R., "Theories for the Interaction of Electromagnetic and Oceanic Waves: A Review", Boundary Layer Meteorol., 13, pp.612-685, 1978.

Wentz, F.J., S.Peteherych, and L.A.Thomas, "A Model Function for Ocean Radar Cross Section at 14.6 GHz", J.Geophys. Res., 89, pp.3689-3704, 1984.

Wentz, F.J., and D.K.Smith, "A Model Function for the Ocean Normalized Cross Section at 14 GHz Derived from NSCAT Observations", J.Geophys. Res., 104, pp.11499-11514, 1999.

Maurizio Migliaccio was born and educated in Italy. He is Associate Professor of Electromagnetics at the University of Cagliari, Sardinia (Italy). He is actually teaching Microwave Remote Sensing at the Faculty of Engineering, University of Cagliari and the Istituto Universitario Navale, Napoli. He has lectured at national and international Universities and research centers. He was guest scientist at the Deutsche Forschungsanstalt fur Lüft und Raumfahrt (DLR), Oberpfaffenhofen, Germany. He is an AGU and AIT member. He is an IEEE Senior Member.

Sea Surface Characterization by Combined Data

ROSALIA SANTOLERI, BRUNO BUONGIORNO NARDELLI, VIVA BANZON(*)
Istituto di Scienze dell'Atmosfera e del Clima-C.N.R, Rome – Italy.
(*) presently at Rosenstiel School of Marine and Atmospheric Sciences, University of Miami, Florida - USA

Key words: altimeter, ocean color, sea surface temperature, Mediterranean

Abstract: In this note the capability of the combined analysis of different satellite data to study ocean dynamic is shown by reviewing some recent results found in various areas of the Mediterranean Sea. ERS/Topex-Poseidon altimeter data, NOAA-AVHRR thermal data and SeaWiFS colour imagery are used to describe the sub-basin and mesoscale dynamic features of the Mediterranean circulation. The resulting picture of the surface circulation obtained satellite observations are compared with concurrent *in situ* and modelled data.

1. INTRODUCTION

Remote-sensing satellites are providing a huge quantity of high quality synoptic data, that represent a powerful and interesting tool for investigation of ocean phenomena at space and time scales not attainable with traditional in situ instrumentation. As a consequence, after a first period during which the possibilities of space oceanography were investigated concentrating on the products obtainable from single sensors, in the last years there has been an effort to combine as much as possible the information obtained from different satellite platforms and sensors and to compare and integrate them with the data collected in situ and obtained from numerical models.

In this section some of the recent results found in various areas of the Mediterranean sea will be presented. The main objectives of this work have been to verify how satellite (ERS/Topex-Poseidon, NOAA-AVHRR thermal data and SeaWiFS colour imagery) data can be used to describe the sub-basin and mesoscale dynamic features of the Mediterranean circulation and to compare the resulting picture of the surface circulation with in situ (LIWEX, SYMPLEX surveys) and modelled data (Iudicone et al. 1998; Marullo et al. 1999; Buongiorno Nardelli et al. 1999, Banzon et al., 2001).

Satellite altimeters directly measure the sea surface slope associated with the oceanographic dynamics. Altimeters data have already proved to be able to resolve sea surface variability in regions corresponding to the major western boundary currents (Gulf Stream, Kuroshio, Malvinas and Angulhas) and the Antarctic Circumpolar Current, moreover direct correlation between sea elevation sensed by satellites and actual dynamic of water masses has been shown for those regions. On the contrary very few works have been done on studies of ocean areas characterised by little sea level slopes or on small semi-enclosed seas like the Mediterranean. This lack of research effort could be probably due to the poor precision of radar altimeters operating before ERS and T/P.

The analysis of altimeter data over the Mediterranean has proved to be useful to study the variability of the Mediterranean circulation on the global scale and identify its main characteristics. Recently Larnicol et al.(1995) and Iudicone et al.(1998) have analysed T/P data over the Mediterranean and found that most of the large scale structure of the circulation have a signature on T/P data. Successively, Ayoub et al. (1998) combined ERS 1 and T/P data to obtain a more complete description of the Mediterranean variable surface circulation.

Although producing high quality data, altimeters provide information only along single lines (groundtracks),. The use of contemporary AVHRR SST (Sea Surface Temperature) data is therefore necessary in order to extend, when it is possible, the one-dimensional altimeter picture of oceanic features to the two-dimensions by using the SST field. An application of the altimeter-SST combination for the Algerian basin is presented in the next paragraph.

In order to perform a correct use of satellite data, a validation by a direct comparison with in situ has been indeed necessary. Moreover satellite-originated informations are related to the surface signature of three-dimensional structures, revealed only by in situ measurements. Combining both, we take advantage of the synopticity of satellites with the completeness of hydrology. This next comparison has been conducted for two different areas, corresponding to the two surveys LIWEX (Levantine basin) and SYMPLEX (Channel of Sicily).

Satellite remote sensing of ocean colour is being increasingly used to better understand the dynamics and the length scales of the biological structures, because of the synopticity and the coverage of space and time scales (e.g. Banzon et al., 2001). In particular, the Sea-viewing Wide Field-

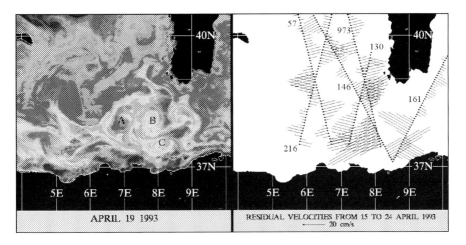

Figure 1: AVHRR image of April 19, 1993 (Ch., 4, NOAA 11)(on the left). Vectors for the estimated geostrophic velocity orthogonal to the ERS-1 and T/P tracks (on the right).

of-view Sensor (Sea Wifs) mission, started in the 1997 provided the oceanography community with a set of high-quality ocean colour data. One of the purpose of the SeaWifs sensor is to supply estimates of the near-surface concentration of phytoplankton pigments by measuring the radiance back-scattered out of the water (LWN) in five different spectral bands (412, 443, 490, 510, 555 nm), thus giving fundamental information on biological near-surface processes but also on physical dynamics at the air-sea interface.

In the last section the usefulness of satellite images as an alternative, or at least as a complement to hydrographic cruises for detecting the occurrence of DWF is pointed out, comparing the patterns of chlorophyll associated to the formation of Deep Water in the Southern Adriatic with the evolution of the surface fields obtained from a numerical model by Levy et al. (1999,2000).

2. THE ALGERIAN BASIN

The Algerian basin dynamics was investigated superimposing along track sea level anomalies and corresponding residual geostrophic velocities to the AVHRR SST images. This allows to better understand links between SLA derived from T/P and ERS-1 data and dynamical features of the Mediterranean circulation. In Fig.1 (NOAA-AVHRR image of 19 April 1993)we show the most representative example of this comparison. The track analyzed are the 146 and 161 (cycle 21,15 April) from T/P and the 973 (15 April), the 57 (18 April), the 130 (21 April) and the 216 (24 April), all during cycle 11 of ERS-1. In April, the Algerian basin images reveal the appearance of a very complex feature in the south-eastern part of the basin.

The image of 19 April 1993 shows an organized structure covering the area between 6.5° E - 8.5° E and 37.2° N - 38.5° N (fig. 3). This area is delimited south by the African coast and east by the rising of bottom topography in the Sardinian channel. This structure seems strictly connected with the eastward flowing Algerian Current, constituted by MAW. It has the shape of three highly convoluted spiral eddies, two of them zonally aligned (A centered at 38.0°N 7.0°E and B centered at 38.1°N 7.9°E respectively) and the third one (C centered in 37.5°N 8.1°E) located South-East of the first two. The eddies have an almost uniform SST of 14.7°C with some colder patches of entrained water. Their mean radius is about 30 km, even if A and C are elliptical. Their spiraling shapes suggest a cyclonic circulation for A and C and an anti-cyclonic circulation for B. The system appears as two coupled mushrooms sharing the middle eddy (B). East of this structure, the eastward MAW flux seems to be limited to a narrow band close the coast.

SLA along T/P track 146 shows a minimum of ~12 cm in correspondence of the C eddy and a maximum of 14 cm for the B one (see fig 2 for the SLA profile and fig.1 for the ground-tracks). The same characteristic is also detected in the SLA along ERS-1 track 973, that has a maximum of 18 cm in correspondence of the B core. The top of the anticyclonic doming have not a regular shape. This could be the result of the entertainment of outer water, as seen in the image. Along the T/P 146 and ERS-1 973 tracks the maximum velocity is between B and C suggesting a westward jet-like flow of ~ 50 cm/s. The idea of the westward flux is validated also from the analysis of velocity along the others westward ERS-1tracks. In particular they suggest that the flux follows the semi-circular structure westward of the tripole. The velocity along ERS-1 130 track, that pass between the two structure A and B, is slower than 973. This could suggests a northward direction of the velocity field in this area, as seems from SST map. No one track give us information on the versus of rotation of A, while they indicate an cyclonic circulation of the tripole. The T/P 161 track is important to characterize the eastward boundary of the structure , the only velocity truly different from zero is near the African coast. It is remarkable that ERS-1 57 and 216 tracks contribute to describe a zone where T/P do not give any information. SLA along these tracks (fig.2) show a maximum near the zone of the core displaced at 39.1 N and 9.3 E. The velocity , about 30 cm/s, reveals that the zone is interested by an anticyclonic circulation with an odd cold core.

Sea surface characterization

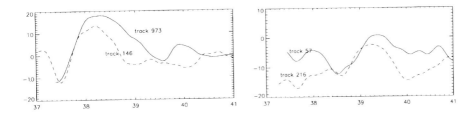

Figure 2: SLA along T/P and ERS-1 ground tracks

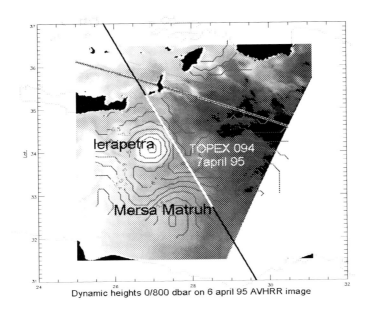

Figure 3. Dynamic heights contours plotted over an infrared AVHRR (ch-4) image in the levantine basin. In white the portion of T/P track 094 showed in fig. 5.

3. ALTIMETER AND AVHRR DATA COMBINED WITH IN SITU LIWEX DATA

The next area under study was the site of the Levantine Intermediate Water EXperiment (LIWEX). This experiment was part of the POEM-BC programme (Physical Oceanography of the Eastern Mediterranean-Biology and Chemistry) and has been carried out as a multinational collaborative effort in the period December 1994-april 1995 in order to study the various

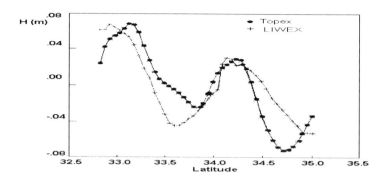

Figure 4. Sea level anomalies (7april) and dynamic heights along T/P track 094 from LIWEX.

aspects of intermediate and deep water formation and spreading in the Levantine basin. The in situ data used in this work are those collected by R/V Urania from march 17 to april 20, 1995. CTD and XBT measurements were taken over an almost-regular grid of station covering a wide region south of Rhodes island, where the previous mentioned gyres are found.

Although LIWEX has not been designed for altimeters validation activities, it was possible to compare data collected during this experiment with TOPEX/POSEIDON measures of sea surface elevation

The ship's data are not properly simultaneous to the passage of the satellite and were not exactly collected along the altimeter tracks. Anyway the result is satisfactory: T/P catches very well the signal associated to the Ierapetra and Mersa Matruh gyres (fig.3), recognisable as maxima in both SLA and dynamic heights referenced to 800 dbar surface (fig.4).

These structures are quite deep, as revealed by temperature, salinity and density fields, reaching more than 600 m depth.

Altimeter validation during SYMPLEX-1 survey (T/P-ERS-AVHRR-CTD-XBT)

The Synoptic Mesoscale Plancton Experiment (ERS-SYMPLEX) has been designed within two international projects approved by ESA (ERS-2 I-110) and by NASA (Symplex) with the aim of studying the relation between the surface signatures visible by remote-sensing satellites and physical and biological internal sea dynamics. This experiment consisted of 4 surveys in the years 1995-1999, and took place in the Channel of Sicily. In this paragraph some of the results of the first SYMPLEX survey will be presented.

During SYMPLEX 1, the number and position of the CTD and XBT casts (done from 12th of April until 13th of May 1995) has been decided in order to have a dense sampling in a region where biological activity is supposed to be very high (a cold filament-like structure at the eastern tip of Sicily) and even a global coverage of the central and eastern regions of the channel. Simultaneous measurements have been conducted during one passage of ERS-2 and two passages of T/P along the selected tracks covering the area.

Owing to its central position in the Mediterranean sea the channel of Sicily controls the large scale thermohaline circulation, i.e. the exchange of water and salt between the eastern and the western basins. It is well known that this area can be considered substantially a two-layer system, separated by an intermediate transition layer (30-100m thick), with fresher water of atlantic origin (modified atlantic water, MAW) in the upper layer, and more saline levantine intermediate water (LIW) in the lower one. From the CTD data analysis results that in spring 96 the MAW stream (identified by lower salinity values) bifurcates at 13.6°E 36.3°N into two branches, one following the sicilian coast and the other flowing offshore, with an intrusion of warm and salty water from the region south of Malta. The northern stream then meets ionian surface waters on the shelf south-east of Sicily, thus resulting in a steep density front.

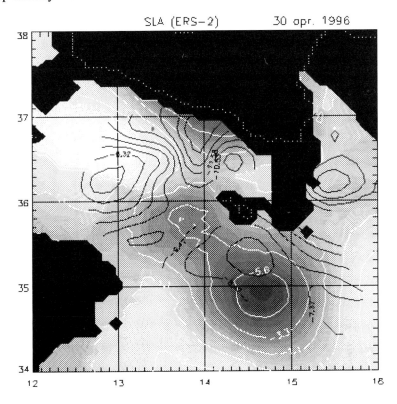

Figure 5. ERS-2 SLA map and dynamic heights contours from SYMPLEX.

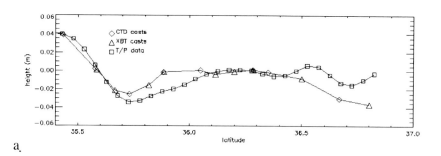

a.

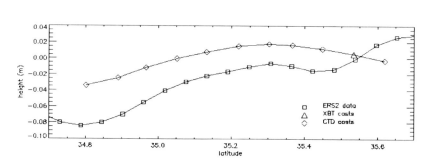

b

c

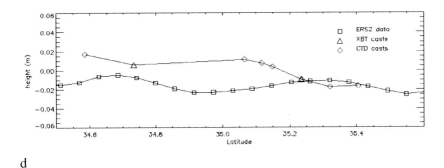

d

Figure 6. Sea level anomalies and dynamic heights along T/P 059 (16 april (a); 6 may (b)), ERS-2 079 (5 may (c)) and 043 (2 may (d)) tracks.

The data collected have been used to compute the dynamic heights of various levels. These calculations provide baroclinic geostrophic velocity only if the reference pressure chosen is parallel to the geoid. In general a no-motion level is used as reference surface. Here we used 300 m as a reference level to get a better coverage of the surface stronger dynamics.

Dynamic heights contours have been compared with an ERS-2 sea level anomaly map computed using optimal interpolation in time and space (fig.5). Both signals clearly reveal the bifurcation of the MAW stream in the central region of the channel. The discrepancy observed east of the sicilian shelf, where the in situ data show a strong minimum not seen in the satellite data, could be related to the removal of the mean from altimeter measurements. In that way the density front which characterises this region would no more contribute to the sea surface elevation.

A better comparison is possible looking at the single tracks (figure 6). In general it can be noted a substantial agreement between satellite and CTD-XBT data. The only differences are found on T/P track 059 (fig.6b) around 36.8°N in may, where the altimeter presents a sloped surface elevation while dynamic heights are flat. The signal resulted very weak along ERS-2 track 043, which is then not to consider with particular interest (fig.6 d).

Residual geostrophic velocities have been calculated from both data sets. Drawing the vectors on the channel 4 AVHRR image (fig.7), we can see that T/P altimeter has caught an eddy at 35.7°N, developed in the surface atlantic water flow and thus bounded by the strongly stratified interface layer. This can be easily observed in the vertical section of velocity calculated from CTD data through geostrophic approximation (fig.8).

The main MAW flow is well identified in the northern part of the channel, while the secondary branch recirculates at about 35.7°N.

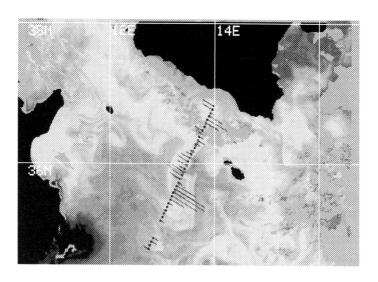

Figure 7. NOAA-14 ch.4 image (17 april) with T/P track 059 velocities (16 april).

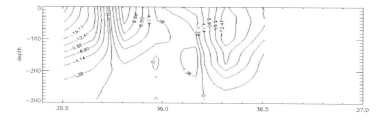

Figure 8. Geostrophic velocity section along T/P track 059.

4. THE ADRIATIC SEA

The Southern Adriatic Gyre is the principal site of Adriatic Deep Water formation (Artegiani et al., 1997), competing with the Cretan Deep Water from the Aegean Sea as principal source of Eastern Mediterranean Deep Water (Roether et al, 1998). Atmospherically-driven convective events are known to affect the water column stratification reflecting on phytoplankton bloom progression. In addition, a process-oriented study by Lévy et al. (1999) showed that following the deep convection phase, the development of the spring bloom continues to be affected by physical processes.

Ocean colour imagery can be used to make inferences about the physical convective process, and to provide time series to elucidate the link between meteorological forcing and chlorophyll distributions (Banzon et al., 2001). It can be shown that, in an open ocean deep convection site, the surface bloom can be shown that, in an open ocean deep convection site, the surface bloom progression can be explained principally by mixed layer deepening and restratification. In this section, the consistency between the scales and spatial patterns of chlorophyll features observed in SeaWiFS with those of the physical convective features described in theoretical and field observation study in other region is discussed in detail.

Three SeaWiFS chlorophyll images of the Adriatic are discussed and compared with previous theoretical and numerical work on other locations. As a principal interpretative basis for the SeaWiFS time series, the surface chlorophyll and density fields from the Levy et al. (1999) can be used (Banzon et al.,2001). These fields were obtained from the simulation of the spring bloom onset following deep convection in the Gulf of Lion and are represented in Figure 9 (I and II-a-c).

The first image shows generally chlorophyll concentrations in the SAG region (< 0.3 mg/m3; Figure 9 a). In the center a "hole" is present, i.e., a zone of minimal chlorophyll concentrations (0.13 to 0.18 mg/m3), which can easily be distinguished from the immediate periphery (> 0.2 mg/m3). The "hole" is more than 50 km across, which is consistent with the size of a mixed convective patch (Marshall and Schott, 1999). Individual plumes cannot be distinguished because they are at the limit of the image resolution (1 km), but the non-homogeneous nature of the mixed patch is quite evident. The beginnings of eddy-like structures are visible at the meandering edge of the "hole".

The pattern resembles the early density distribution in the Levy et al. (1999; 2000) simulations, but does not show well in their surface pigment map because of the colour scale interval they use (Figure 9 I-a and II-a). The phytoplankton distribution strongly reflects the surface density field only when restratification develops, as during active convection photosynthesis is inhibited by vertical mixing. As the mixed layer deepens, vertical movements distribute algal cells over a larger vertical extent and decrease the exposure time of individual algal cells to the light. These two factors cause chlorophyll concentrations to drop below that of an already low background level.

Figure 9 b shows the contraction of the minimal-chlorophyll "hole" and the development of series of meanders at its edges. Highlighted by the steep gradient in pigment concentrations, these meanders are in the order of tens of km, consistent with the observed size of laterally spreading eddies in the Mediterranean. As described by Marshall and Schott (1999) and references therein, during the lateral exchange phase, mesoscale baroclinic instabilities appear on the surface density field in the form of meanders at the outer edge of the convective region, as shown by the Levy et al. (2000) simulation (Figure 9 I-b and II-b).

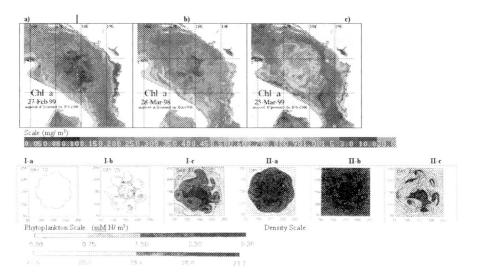

Figure 9: Principal phases of the South Adriatic phytoplankton bloom onset following deep convection: a) chlorophyll minimum feature at the end of the deep convection phase; b) contracted chlorophyll minimum with enhanced chlorophyll at the meandering edges; c) final phase of the bloom onset: well developed spring bloom. For comparison, selected chlorophyll (I-a, I-b, I-c) and density (II-a, II-b, II-c) fields produced by simulations using real meteorological forcing in the Gulf of Lion are shown (adapted from Levy et al., 2000).

Mesoscale upward motions are responsible for shoaling of the mixed layer in the trough of the meanders; consequently in the areas with a shallow mixed layer, the exposure time of the phytoplankton cells to the light is increased and a crest-like phytoplankton distribution shows in the model results (Levy et al., 1999).

The maximum in chlorophyll concentrations at the border of the "hole" in Figure 9b is consistent with the model and field data in the Gulf of Lion. This crest-like distribution was not observed in CZCS data by Levy et al. (1999), thereby limited their model validation. Such a pattern has been verified using satellite ocean colour data for the first time by Banzon et al. (2001). The success in observing this distribution is due to the higher spatial and temporal sampling resolution and the greater instrumental sensitivity of SeaWiFS relative to CZCS.

The third image exemplifies a well-developed but asymmetrical bloom with chlorophyll concentrations exceeding 2.0 mg/m3 in the northern and eastern part of the SAG (Figure 9c). The structures are made up of smaller filamentous patches that reflect the mesoscale instabilities that have developed from the meanders. The image shows a chlorophyll distribution similar to the model (Figure 9 I-c) where the minimum-chlorophyll central "hole" has completely disappeared.

The strong similarities observed between the Levy et al. (1999, 2000) model results for the Gulf of Lion and the individual SAG SeaWiFS images suggest that the underlying theoretical concepts relating surface pigment distributions and convective processes are valid. Certainly, an exact correspondence cannot expected since the forcing is usually simplified in the numerical experiments, except for the one case when real meteorological data specific to the Gulf of Lion was used (Levy et al., 2000). Nevertheless, the generality applicability of the concepts embodied in the model is demonstrated by our ability to observe in the SAG region all the main features characteristic of the spring bloom evolution following deep convection. Thus, despite complex winds, highly variable heat forcing, and their interaction with local circulation, this approach may be extended easily to satellite imagery of other deep convection sites

5. BIBLIOGRAPHY

Artegiani, A., D. Bregant, E. Paschini, N. Pinardi, F. Raicich, A. Russo. The Adriatic Sea general circulation: Part I: Air-Sea Interactions and water mass structure, *J. Phys. Oceanogr.* 27, 1492-1514, 1997.

Ayoub, N., P.Y. Le Traon, and P. De Mey, A description of the Mediterranean surface variable circulation from combined ERS-1 and Topex/Poseidon altimetric data, *J. Mar. Syst.*, 18(1-3), 3-40, 1998.

Banzon V, S. Marullo, F. D'Ortenzio, R. Santoleri and R. Evans, SeaWiFS observations of the surface phytoplankton bloom in southern Adriatic Sea (1998-2000): year-to-year variability and the role of atmospheric forcing and deep convection, submitted to *J. Geophys. Res.*, 2001.

Blaha, J., and B. Lunde, Calibrating altimetry to geopotential anomaly and isotherm depths in the western north Atlantic, *J. Geophys. Res.*, 97, 7465-7477, 1992.

Bouzinac, C., J. Vazquez, and J. Font, CEOF analysis of ERS-1 and TOPEX/POSEIDON combined altimetric data in the region of the Algerian current, *J. Geophys. Res*, 103, 8059-8071, 1998.

Buongiorno Nardelli B., Santoleri R., Marullo S., Iudicone D., Zoffoli S., 1999. Altimetric signal and three dimensional structure of the sea in the Channel of Sicily, *J. Geophy. Res.*, 104, 20585-20603.

Buongiorno Nardelli B., Iudicone D., Santoleri R., Zoffoli S., 1997. Mediterranean sea level anomalies and three dimensional structure of the sea from LIWEX and Symplex, Proceedings 3 Ers Symposium' Space at the service of our enviroment', Florence, march 1997, ESA, SP-414, vol.II, 1467-1473.

Challenor, P.G., J.F. Read, R.T. Pollard, and R.T. Tokmakian, Measuring surface currents in the Drake passage from altimetry and hydrography, *J. Phys. Oceanogr.*, 26(12), 2748-2759, 1996.

Hernadez, F., and P.-Y. Le Traon, Mapping mesoscale variability of the Azores current using TOPEX/PoseidoN and ERS 1 altimetry, together with hydrography and lagrangian measurements, *J. Geophys. Res*, 100, 24995-25006, 1995.

Iudicone, D., S. Marullo, R. Santoleri, and P. Gerosa, Sea level variability and surface eddy statistics in the Mediterranean sea from Topex/Poseidon data, *J. Geophys. Res.*, 103, 2995-3012, 1998.

Larnicol, G., P.Y. Le Traon, N. Ayoub, and P. De Mey, Mean sea level and surface circulation variability of the Mediterranean Sea from 2 years of TOPEX/POSEIDON altimetry, *J. Geophys. Res.*, 100, 163-177, 1995.

Le Traon, P.Y., and P. De Mey, The eddy field associated with the Azores Front east of the Mid-Atlantic Ridge as observed by the GEOSAT altimeter, J. Geophys. Res., 99, 9907-9923, 1994. LaViolette, P.E., J. Tintore and J. Font, The surface circulation of the Balearic Sea, *J. Geophys. Res.*, 95, 1559-1568, 1990.

Lévy, M.., L. Mémery, G. Madec, The onset of the spring bloom in the MEDOC area: mesoscale spatial variability. *Deep Sea Res.*, 46, 1137-1160, 1999.

Levy, M.., L. Mémery, G. Madec, Combined effects of mesoscale processes and atmospheric high-frequency variability on the spring bloom in the MEDOC area. *Deep Sea Res.*, 47, 27-53, 2000.

Le Vourch, J., C. Millot, N. Castagne, P. Le Borgne and J.P. Orly, Atlas of thermal fronts of the Mediterranean Sea derived from satellite imagery. 16, Mem.Inst.Oceanogr.of Monaco, 1992.

Marshall, J and F. Schott (1999) Open ocean deep convection: observations, models and theory, *Rev. Geophys.*, 37, 1, 1-64

Marullo S., R. Santoleri, P. Malanotte Rizzoli and A. Bergamasco, 1999: The sea surface temperature field in the eastern Mediterranean from AVHRR data. Part I. Seasonal variability, *J. Mar. Sys.*, 20 (1-4), 63-81.

Marullo, S., R. Santoleri, P. Malanotte Rizzoli, and A. Bergamasco, 1999: The sea surface temperature field in the eastern Mediterranean from AVHRR data. Part II. Interannual variability, *J. Mar. Syst.*, 20(1-4), 83-112,.

Marullo S, Zoffoli S. Santoleri R., Iudicone D., Buongiorno Nardelli B., 1997. A combined use of ERS-1 and Topex/Posidon data to study the mesoscale dynamics in the Mediterranean Sea, Proceedings 3 Ers Symposium' Space at the service of our enviroment', Florence, march 1997, ESA, SP-414, vol.II, 1485-1491.

Roether, W., B. B. Manca, B. Klein, D. Bregant, D. Georgopoulos, V. Beitzel, V. Kovacevic, and A. Luchetta, Recent changes in Eastern Mediterranean deep waters, *Science*, 271, 333-335, 1996.

Santoleri R., Buongiorno Nardelli B., Iudicone D., Zoffoli S., Marullo S., 1997. Combined use of Ers, Topex/poseidon, AVHRR and in situ data to study the mesoscale dynamics in the Mediterranean Sea, Proceedings European Symposium on Aerospace Remote Sensing, London 1997, SPIE, vol.3222, 400-411.

Strub, P.T., T.K. Chereskin, P.P Niiler, C. James, and M.D. Levine, Altimeter-derived variability of surface velocities in the California Current System, 1, Evaluation of TOPEX altimeter velocity resolution, *J. Geophys. Res.*, 102, 12727-12748, 1997.

Chapter 4

Data Assimilation and Remote Sensing Programs

Mesoscale Modeling and Methods Applications

ROSSELLA FERRETTI
Department of Physic, University of L'Aquila.

Key words: Data Assimilation, Objective Analysis, Nudging Technique, Meso-scale modelling

Abstract: A review of Variational Data Assimilation together with a brief historical review is presented. Furthermore, a few methods used for improving Initial and Boundary Conditions in Numerical Weather Prediction are presented. The application of simple techniques applied to a mesoscale model, to improve the local initial condition, is presented . Both model aided studies and numerical experiments performed using the MM5 show that the assimilation of conventional data at high resolution may improve the forecast of the precipitation and the understanding of the local circulation.

1. INTRODUCTION

The rapid growth of computer power allows the development of increasingly fine resolution numerical weather prediction models improving both our understanding of atmospheric processes and the forecast of severe events. However, the higher resolution of the numerical models alone does not produce an equal gain in the forecast skill. In the last years, it became clear that both the Initial Conditions and the physical processes (convection, turbulence and radiation) play an important role in the weather forecast. This lecture will focus on the Initial Conditions.

This is a largely studied topic since the first half of the nineteenth century, at that time a few attempts were made to represent meteorological observations in a regular grid. Indeed, Bjerknes was convinced that weather prediction was an Initial-Value problem in which by solving the governing equation the future atmospheric states would be known. Richardson (1922) performed the first attempt to initialize a model governing equations, written in finite-differences form, and integrated forward in time. At initial time the pressure was defined on a regular grid together with the wind velocities. To have the observing values on a regular grid Richardson used a subjective analysis. Unfortunately, this experiment failed, one of the reasons was the Initialization which was not understood at that time. To get a successful weather forecast we have to wait until 1950, using both the Bjerknes and Richardson ideas a forecast was performed at Princeton Institute (Charney et al., 1950 and Platzman, 1979). At that time, the procedure to estimate the observed values, at a recording station on a regular grid, was still subjective. The first attempt of Objective Analysis was done by Panofsky (1949), he used a polynomial expansion to fit all the observations in a small area. The coefficients of the expansion were produced using a least square fit. A step forward to this procedure was produced by Gilchrist and Cressman (1954) whose introduced the concept of region of influence. Further improvements were introduced to date, for a good historical review see Daley (1991).

In the following sections a brief description of a few methods currently used to produce the daily best Initial Conditions for the high resolution weather forecast is presented. A few examples of sensitivity study to improve the high resolution numerical weather prediction will be shown. Finally, two methods for improving either the mesoscale Initial Conditions or the model parameter run time, using the four dimensional data assimilation, will be presented, and a few applications will be discussed.

2. VARIATIONAL ASSIMILATION

In this section a brief review of the state of the art of the methods used for producing the best Initial Conditions for the weather forecast is presented.

A good review of the data assimilation methods and theories has been given by Lorenc (1986) and Ghil and Malanotte-Rizzoli (1993). Furthermore, an excellent review of the initializations methods and a presentation of the several data analysis technique can be found in Daley's book (1991).

Recently the assimilation schemes are formulated *variationally*, that is essentially the analysis is determined as that atmospheric state that minimizes a measure of a fit between it and both the background data and the observations. The fit is formulated using weights which both depend on the error statistics of the data and include constraints.

If only a column or a sounding is analyzed by the variational method, the 1D-VAR is applied. If only a surface is analyzed by the variational method at a single time then the 2D-VAR is applied, obviously if a volume is

Mesoscale modeling and methods applications

analyzed by the variational method at a single time then the 3D-VAR is applied. Finally, when the data are variationally fitted over a period of time using a forecast model as a strong constrain then the 4D-VAR is applied.

For large scale models the current synoptic scale for the radiosonde is sufficient to reproduce a reasonable representation of the atmosphere at the model resolution, whereas for mesoscale models the observations are much less than the degree of the model freedom. Furthermore, the rapid growth of new types of data generates the problem of how to incorporate them into the models, and, as already pointed out, the rapid increase of computer power allows for high resolution forecasts which turns into a need for more observations.

A simple introduction to 4D-VAR is presented using an idealized method based on Bayesian probabilistic arguments by Lorenc (1986). The Bayes' theorem applied to our problem states that a posterior probability of an event x_t occurring, given that event y_0 is known to have occurred is proportional to the prior probability of x_t multiplied by the probability of y_0 occurring given that x_t is known to have occurred:

$$P_a(x) = P(x_t/y_0) = P(y_0/x_t)\, P(x_t)$$

where the subscript t indicates the true value, and o the observed one. P_a is a Probability Distribution Function and the best estimates of x_a is given by the minimum variance:

$$x_a = \int x\, P_a(x)\, dx$$

or by the maximum likelihood:

$$x_a = x \text{ such that } P_a(x) \text{ minimum}$$

the accuracy of x_a is a necessary information that is contained in $P_a(x)$.

To proceed we need to specify the Probability Distribution Functions (PDF) : if they are assumed Gaussian and we use the maximum likelihood we need to minimize the following operator:

$$J = \{y_0 - k_n(x)\}^T (O + F)^{-1} \{y_0 - k_n(x)\} + (x - x_b)^T B^{-1} (x - x_b).$$

where for these PDFs **O**, **B**, and **F** are covariance matrices.

The 4D-VAR (Rabier et al., 1998) requires to minimize the following operator:

$$J(\delta x) = 0.5\, \delta x^T B^{-1} \delta x + 0.5\, \Sigma_i (H_i\, \delta x(t_i) - d_i)^T R^{-1} (H_i\, \delta x(t_i) - d_i)$$

where **B** is the background error covariance, **R** the observation error covariance, and H_i is a linear approximation at time t_i of the observation operator H_i. d_i is the innovation vector given by:

$$d_i = y_i^0 - H_i \, x^b(t_i)$$

where $x^b(t_i)$ is computed by integrating a non-linear model M from time t_0 to time t_1 starting from x^b, i is the time index, and δx is the increment to be added to the background x^b at the starting point i=0, that is:

$$\delta x(t_i) = M(t_i, t_0) \, \delta x$$

where **M** is a tangent-linear model used to propagate the increments in time (the ECMWF model runs at lower resolution).

The minimum of the function J is computed using minimization methods:1) the quasi-Newton, 2) a conjugate gradient (first derivative). The gradient is computed using the adjoint model.

Until 1997 Optimal Interpolation has been used operationally at European Center for Medium Range Weather Forecast (ECMWF), the 4D-VAR replaced the 3D-VAR system (it was operational from 30 Jan 1996) at ECMWF from 25 Nov 1997, (Courtier et al., 1998; Rabier et al., 1998; Andersson et al., 1998 and Rabier et al, 1998).

3. THE MM5 ADJOINT MODEL

Briefly an example of an adjoint model is presented in the following paragraph using the MM5. The adjoint of the MM5 can be symbolically written as (Zou et al., 1995)

$$X(t_r) = P_r(x) \begin{pmatrix} x_0 \\ b \end{pmatrix}$$

where x_0 is any model variable: u, v, T, q, p', w $|_{t=t_0}$ for the Initial Conditions, **b** are the Boundary Conditions, $x(t_r)$ is any model forecast variable at time t_r and P_r represents all the operator matrices in the MM5.

The <u>tangent-linear</u> model can be written as follows:

$$X'(t_r) = P'_r(x) \begin{pmatrix} x_0 \\ b \end{pmatrix}$$

Where **x'** is a perturbed model solution and **P'** is the tangent-linear model operator.

Finally the adjoint model

$$\begin{pmatrix} x_0 \\ b \end{pmatrix} = [P_r(x)]^T x(t_r)$$

where **x** is the adjoint variables, and $(\mathbf{P})^T$ is the adjoint model operator.

The 4DVAR is a computing time expensive technique, therefore to improve the high resolution weather forecast a different technique is now presented.

4. THE HIGH RESOLUTION NUMERICAL WEATHER FORECAST

During the last years, in the framework of a cooperation between Parco Scientifico and Tecnologico Abruzzo (PSTA) and University of l'Aquila - Department of Physics (UNIAQ) the MM5 (Grell et al., 1994) was run operationally at high resolution (3km) over Abruzzo region (Italy).This is a complex system made by :
-Pre-processing
-Forecast
-Post-processing

The pre-processing takes care of the interpolation of the Initial Conditions from the meso-alpha (which approximately corresponds to a scale of motion in which the mass adjustment is performed by quasi – geostrophic motion) to the mesobeta-scale (which approximately corresponds to a scale of motion in which vertical accelerations become important) using ECMWF data analyses. A bilinear interpolation is used associated with a Lambertian conformal projection, that is the model grid. The MM5 (PSU/NCAR) is used for the very forecast, it is a fully compressible model at the primitive equation, the non-hydrostatic version is used. Several parameterization schemes are available for the Planetary Boundary Layer, the radiation, the microphysics and the cumulus convection. The ones used operationally are briefly described in Paolucci et al., (1999). 24 vertical levels unequally spaced are used and the vertical coordinate is σ ,a terrain following coordinate. Three domains two way nested are used to allow the high resolution over the Abruzzo region. The mother domain (centered over southern Europe) has a grid size of 27km, the second domain has a grid size of 9km and it is centered over central Italy, finally the innermost domain, having a grid size of 3km, is centered over Abruzzo, a region in the central-east of Italy.

A few sensitivity tests to explore the skill of the MM5 in reproducing thermally driven circulation were performed (Mastrantonio et al., 2000). During Feb, 1996 a winter sea breeze was observed by a SODAR along the west coast close to Rome. Several model simulations were performed to asses the winter sea breeze, to study the role played by the sea surface temperature and to verify the model results by comparison with the SODAR data. The model set up is the same as the one used operationally at PSTA/UNIAQ. As verification of the sea breeze event, a simulation using a thick layer of clouds was performed, to prevent the thermally driven circulation. The model clearly showed no sea breeze if the clouds were used. Furthermore, the observed sea surface temperature allowed for a stronger sea breeze development suggesting that this is an important parameter at high resolution for a correct simulation of the local circulation (Ferretti et al., 2002).

An assimilation system of satellite imagery data was tested to simulate thermally forced circulation by Ruggiero et al.(2000).The experimental set up was different from the one previously discussed , they used the MM4 and the assimilation of satellite data was performed every 6h associated with conventional observations to drive the model simulations. The response of the model (MM4) was good, the model promptly returned to the normal state, furthermore the data assimilation allowed to improve the representation of the surface insulation.

To the aim of improving the weather forecast a few experiments for generating the best mesoscale Initial Conditions using both conventional data and asynoptic data runtime are now presented.

5. ASSIMILATION TECHNIQUES

5.1 Objective Analysis

To improve the Initial Condition for the high resolution weather forecast the re-assimilation of conventional data is tested. The ***Objective Analysis*** of observations blended together with ECMWF data analyses, which are used as First Guess, allows to modify the data analyses itself. Two different methods were tested (Faccani et al., 2001):

1) Cressman (circular and banana shaped) (CRS)
2) Multiquadric (MQD)

The Cressman (Cressman, 1959) modifies the grid points enclosed into a circular or banana shaped area centered at the station using:

$$\alpha' = \alpha_{i,j} + \left(\frac{\sum_{k=1}^{N} w_{(i,j),k} D_k}{\sum_{k=1}^{N} w_{(i,j),k}} \right)$$

where α' is the analysis at i, j point, N is the total number of stations, D_k is the differences between the k-the observation and the analysis at the station location and $w_{(i,j),k}$ is the weight that is given by the following function:

$$w_{(i,j),k} = (R^2 - d^2_{i,j}) / (R^2 + d^2_{i,j})$$

where R is the radius of a circular area, and $d_{i,j}$ is the distance between the grid point and the k-th station.

The Multiquadric MQD (Nuss and Titley, 1994) allows to modify the grid points considering the whole set of observations, furthermore it accounts for statistic error of the measurements. The function used to compute the analysis is the following:

$$\alpha' = \alpha_{i,j} + \mathbf{Q}_{(i,j),k} [\mathbf{Q}_{k,m} + (N \lambda \sigma^2_k \delta_{k,m})]^{-1} \mathbf{D}_m$$

where $\mathbf{Q}_{(i,j),k}$ is the matrix of the radial function depending on the distance between the observation and the grid point, $\mathbf{Q}_{k,m}$ is the matrix of the radial function, $\mathbf{D}_m$ is the interpolation matrix, σ^2_k is the statistical error associated with the meteorological variables, and N are the observations.

The Piedmont flood is used as test case by Faccani et al., (2001) to verify the impact of the re-assimilation of surface and upper air data into the I.C. by Objective Analysis. Previous study using the same model (Ferretti et al., 2000) showed a good skill in the rainfall distribution over the flood area, but over the sea and far from the flood the forecast was poor compared to other models (Richard et al., 1998). Furthermore, this case showed peculiarity into the meteorological characteristics: this event was characterized by a strong large scale forcing producing most of the precipitation which caused the flood, but a local convergence area over the Po Valley produced a strong uplifting of the flow and it held back the frontal system (Rotunno and Ferretti, 2001). The MM5 set up for these experiments was similar to the one used operationally (Paolucci et al., 1999) but the high resolution domain was located over the flood area, 3 domains two way nested respectively at 27km, 9km and 3km were used. The results (Faccani et al., 2001) show an improvement of the rainfall in the Piedmont area allowing for a splitting of the 300mm contour, which turned into a better agreement with the observations.

5.2 Newtonian Relaxation

Another technique was tested to improve the forecast of the precipitation, this is based on the *-Four-Dimensional Data Assimilation (FDDA)*. The FDDA consists in combining current and post data in an explicit dynamical model so that the prognostic equations provide time continuity and dynamic coupling among various fields (Charney et al., 1969). Similarly the **_Nudging technique_** uses a continuous dynamical assimilation, and a forcing function is added to the model equations to gradually nudge the model variables toward the observations (Anthes, 1974; Kuo and Guo, 1974; Stauffer and Saeman, 1990 etc.). The forcing term gradually corrects the model ensuring that the variables remain in balance. The nudging can be applied both to the analysis or the observations. The model governing equations change accordingly for the two cases. If the nudging is applied using the analysis

$$\frac{\partial p^* \alpha}{\partial t} = F(\alpha,x,t) + \overset{1}{G_\alpha W_\alpha(x,t)\varepsilon_\alpha(x)p^*(\hat{\alpha}_0 - \alpha)} +$$
$$\overset{2}{G_p W_p(x,t)\varepsilon_p(x)\alpha(\hat{p}^* - p^*)}$$

then the general flux form for the modeling equations is the following:

where F represents all the physical forcing as the advection term, the Corioli's force etc.., α are the model's dependent variables, x are the independent spatial variables and t is the time. The term 1 is the nudging factor for α, being G the nudging factor and W the weight function. The term 2 is the nudging factor for p, G and W as before, and ε is the quality factor. If the Nudging of the observations is performed the corrections computed at the observation location are applied at the grid in a region surrounding the observations. Assuming $G_p = 0$, the following equation can be written:

$$\frac{\partial p^* \alpha}{\partial t} = F(\alpha,x,t) + G_\alpha p^* \frac{\overset{1}{\sum_{i=1}^{N} W_i^2(x,t)\gamma(\alpha_0 - \hat{\alpha})}}{\sum_{i=1}^{N} w_i(x,t)}$$

where F and G are defined as before, i is the i-th observation of a total of N which are within a radius of a given grid point, α_0 is the locally observed value of α that are the model prognostic variable interpolated at the observation location, γ is the observational quality factor (0, 1) and w is the weighting factor accounting for the spatial and temporal separation of the i-th observation from a given grid point at any time step.

This technique was used to assimilate the precipitable water estimated by the Global Positioning System (GPS) (Pacione et al., 2000a) into a mesoscale model to improve the precipitation forecast (Pacione et al.,2000b).

The computation of the PW using the GPS data is performed by accounting for the delay of the GPS signal from the satellite to the ground caused by the variation of the refraction index because of T, P and q (Pacione et al., 2000a):

$$\Delta L^{Zen} = 10^{-6} \int (k_1 \frac{P}{T} + k_2 \frac{e}{T} + k_3 \frac{e}{T^2}) dz$$

where P, T and e are respectively the atmospheric pressure, temperature and partial pressure of water vapor, and k_i are coefficients that have to be chosen.

The total delay (ZTD) can be separated into an hydrostatic part $\Delta L_h(P)$:

$$\Delta L_h^{Zen} = (2.2768 \pm 0.0024 \times 10^{-7}) \, P/f(\lambda, H)$$

and a wet delay $\Delta L_w(q_v, T)$:

$$\Delta L_w = ZTD - \Delta L_h^{Zen}$$

where λ is the latitude at the station location, H is the station height; the ZTD is given by the GPS using a mapping function to account for the satellite elevation,

Finally the precipitable water can be obtained by the following relationship.

$$PW = \Pi(T_m) \, \Delta L_w^{Zen}$$

where Π is a weak function of the weighted mean temperature.

The GPS Precipitable Water (PW) is not easily assimilated into the MM5, therefore the nudging technique was used (Faccani et al., 2002). The MM5 allows only for temperature, wind and water vapor nudging, therefore the GPS data has to be converted into one of the model variable that is the PW, which was computed by integrating q_v (Kuo et al., 1993) over the model layers:

$$PW = (p^*/g) \, (\Sigma_k \, q(k) \, \Delta\sigma(k))$$

where q(k) is the model specific humidity, $\Delta\sigma(k)$ are the layers thickness of the model, kx is the total number of model layers, p^* is given by $p_s\text{-}p_t$.

The PW_m estimated by using the model q_v is compared with the one obtained by the GPS PW_o, if the ratio between PW_o and PW_m is less than a critical value then q_v is computed by using:

$$q_m^{i+1}(k) = q^i(k)\left(\frac{PW_o}{PW_m}\right)$$

The cycle is closed by computing a new $PW_m.^{i+1}$.

Several experiments were performed to evaluate the impact of this technique and of the OA, previously described, on the high resolution forecast of the precipitation.

To estimate the impact of the GPS data of Cagliari, L'Aquila and Matera into the high resolution precipitation forecast a study was performed. A MAP case was chosen for this experiment, the one occurred during Oct 22, 1999. The meteorological situation of this event was characterized by a surface low pressure associated with a warm and moist advection from south-west over the Mediterranean sea, which was followed by a cold front associated with an upper level trough. The southwesterly flow produced rainfall over most of the Italian regions, the precipitation forecast over the northern regions was good enough, whereas the one over the central-south ones was not good. Therefore, an attempt is made to improve the local precipitation forecast by assimilating the local data. A few simulations were performed: 1) using the assimilation of the conventional data only (OA); 2) without assimilation (CNTR); 3) using the nudging of the PW only (ND_PW); 4) as 3 but associated with the assimilation of conventional data (OA_ND).

The comparison among all the simulations and the observed data suggests that the assimilation of conventional data has a strong impact on the precipitation forecast, whereas the nudging of the PW only is not sufficient to correct the lack of information into the Initial Conditions. Therefore, the OA associated to the nudging technique allows for the best results for this case (Faccani et al., 2002).

6. CONCLUSIONS

A brief review of the most important techniques used to incorporate remote sensing data to generate the best Initial Conditions was presented. The Initial and Boundary Conditions daily produced at the ECMWF, using the 4D-VAR, contain information up to a mesoalpha-scale. They generally allow for a good forecast, but the growth of the computer power allows to reach a finer resolution making these Initial Conditions no more sufficient for a good forecast at that resolution. If a high resolution forecast is performed the information contained into the Initial Conditions are not sufficient to ensure a good forecast. Therefore the lack of operational measurements at the mesobeta-scale is a critical point. In fact in the last years a few sensitivity studies successfully showed the positive impact of the assimilation of the local observations into the weather forecast using the Objective Analysis. To gain information at the mesobeta scale, the assimilation of local data is urgent. At finer resolution it has been shown that even the re-assimilation of conventional data is important for improving the weather forecast.

Furthermore, in the late 90', remote sensing both ground based or space-borne became more or less operational in producing measurements at high resolution. Instruments , such as profilers, sodar, lidar and Doppler radar provide almost continuos information of the evolving atmosphere. Therefore as future work the assimilation of ground based remote sensing may allow to improve the 3D wind with a gain in the estimation of the vertical motion .The latter is an important variable for the non-hydrostatic models.

The assimilation of satellite data may allow to improve the representation of both the local circulation and of the wet parameters of the atmosphere turning into a better forecast of the precipitation.

Recently several studies were focused on the estimation of the PW by the GPS with the aim of assimilating the PW into numerical models by using either variational methods or nudging technique. A further improvement may be achieved by using the Rain Rate estimated from SSM/I.

7. REFERENCES

Andersson, E., Haseler, J., Undén, P., Courtier, P., Kelly, G., Vaisiljevic, D., Brankovic, C., Cardinal, C., Gaffard, C., Hollinsworth, A., Jakob, C., Janssen, P., Klinker, E., Lanzinger, A., Miller, M., Rabier, F, Simmons, A., Strauss, B., Thépaut, J.-N. and Viterbo, P., 1998: The ECMWF implementation of three dimensional variational assimilation (3D-Var) , III: Experimental results. *Q. J .R. Meteorol. Soc.*,**124**, 1831-1860.

Anthes, R. A., 1974: Data assimilation and initialization of hurricane prediction models. *J. Atmos. Sci.,* **31**, 702-719.

Charney, J. G.,Fjortoft, and J. von Neumann, 1950: Numerical integration of the barotropic vorticity equation. *Tellus,***2**, 237-254.

Charney, J. G., M. Halem and R. Jastrow, 1969: Use of incomplete historical data to infer the present state of the atmosphere. *J. Amos. Sci.*, **26**, 1160-1163.

Cressman, G.P., 1959: An operational objective analysis system. *Mon. Wea. Rev.*, **87**, 367-374.

Courtier, P., Andersson, E., Heckley, W., Pailleux, J., Vaisiljevic, D., Hamrud, M., Hollinsworth, A., Rabier, F. and Fisher, M, 1998: The ECMWF implementation of three dimensional variational assimilation (3D-Var), I: formulation. *Q. J .R. Meteorol. Soc.*,**124**, 1783-1807.

Daley, R., 1991: *Atmospheric Data Analysis,* Cambridge University Press, p.457.

Lorenc, A.C., 1986: Analysis methods for numerical weather prediction. *Q. J .R. Meteorol. Soc.*,**112**, 1177-1194.

Faccani, C., R. Ferretti and G. Visconti, 2001: Assimilation of conventional data into the ECMWF analysis to improve the high resolution weather forecast. *Conditionally accepted by Mon. Wea. Rev.*.

Faccani, C., R. Ferretti, R. Pacione, F. Vespe and G. Visconti, 2002: Incorporating the GPS PW using assimilation and nudging techniques in a Limited-Area Model: A case Study for IOP8 MAP/SOP. *To be submitted to J. Geophys. Res.*

Ferretti, R., S. Low-Nam, and R. Rotunno, 2000: Numerical simulations of the Piedmont flood of 4-6 November 1994. *Tellus*, **52**, 162-

Ferretti, R, G. Mastrantonio, A. D'Onofrio, G., A. Viola, and S. Argentini, 2002: A thermally driven circulation during winter on the Mediterranean sea along the Italian Tyrrenian coast. *To be submitted to J. Geophys. Res.*

Ghil, M. and Malanotte-Rizzoli, 1991: Data assimilation in meteorology and oceanography. *Adv. Geophys.*, **33**, 141-265.

Gilchrist, B and G. Cressman, 1954: An experiment in Objective Analysis, *Tellus*, **6**, 309-318.

Grell G. A., J. Dudhia, and D. R. Stauffer, 1994: A description of the fifth-generation Penn State/NCAR Mesoscale Model (MM5). NCAR Tech. Note, NCAR/TN-398+STR, National Center for Atmospheric Research, Boulder, CO, 138 pp.

Kuo, Y.-H., and Y.-R. Guo, 1989: Dynamic initialization using observations from a hypothetical network of profiles. *Mon. Wea. Rev*, **117**, 1975-1998.

Kuo, Y.-H., Y.-R. Guo, and E. R. Westwater, 1993: Assimilation of precipitable water measurements into a mesoscale numerical model. *Mon. Wea. Rev*, **121**, 1215-1238.

Mastrantonio G., A. Viola, S. Argentini, R. Ferretti and A. D'Onofrio., 2000: SODAR contribution to the improvement of the low-level circulation numerical modeling. *Preprint of the Tenth International Symposium on Acoustic Remote Sensing and associated techniques of Atmosphere and Oceans.* Auckland, New Zeland, 2000.

Nuss, W. A. and D. W. Titley, 1994: Use of multiquadric interpolation for meteorological objective analysis. *Mon. Wea. Rev.*, **122**, 1611-1631.

Pacione, R., C. Sciarretta, F. Vespa, C. Faccani and R. Ferretti, 2000a: GPS PW assimilation into the MM5 with the nudging technique. *Phys. Chem. Earth (A)*, **26**. ,489-485.

Pacione R., C. Sciarretta, C. Ferraro, C. Nardi, F. Vespa, C. Faccani, R. Ferretti, E. Fionda, and L. Mureddu, 2000b: GPS meteorology comparison with independent techniques. *Phys. Chem. Earth (A)*, **26**. ,139-145.

Panofsky, H, 1949: Objective weather-map analysis. *J. Appl. Meteor.*,**6**, 386-392.

Paolucci T., L. Bernardini, R. Ferretti and G. Visconti, 1999: MM5 real-time forecast of the catastrophic event on May 5, 1998. *Il Nuovo Cimento*, **22**, 727-736.

Platzman, G., 1979: The ENIAC computations of 1950: Gateway to numerical weather prediction. *Bull. Amer. Meteor. Soc.*, **60**, 302-312.

Rabier, F, McNally, A., Andersson, E., Courtier, P., Undén, P., Eyre, J., Hollinsworth, A., and Bouttier, F, 1998: The ECMWF implementation of three dimensional variational assimilation (3D-Var), II: Structure functions. *Q. J .R. Meteorol. Soc.*,**124**, 1809-1829.

Rabier, F, Thépaut, J.-N. and Courtier, P., 1998: Extended assimilation and forecast experiments with a four-dimensional variational assimilation system. *Q. J .R. Meteorol. Soc.*,**124**, 1861-1887.

Richard, E., J. Charney, S. Cosma, R Benoit, S. Chamberland, A. Buzzi, L. Foschini, R. Ferretti and J. Stein, 1998: Intercomparison of simulated precipitation fields for some MAP episodes. *Map Newsletter*, **9**, 12-13.

Richardson, L, 1922: *Weather prediction by numerical process.* Cambridge University Press.

Rotunno, R., and Ferretti, R., 2001: Mechanisms of intense alpine rainfall. *J. Atmos. Sci.,* **58**, 1732-1749.

Ruggiero F. H., G. D. Modica and A. E. Lipton, 2000: Assimilation of satellite imager data and surface observations to improve analysis of circulations forced by cloud shading contrasts. *Mon. Wea. Rev.*, **128**, 434-448.

Stauffer D. R. and N. L. Seaman, 1990: Use of Four-Dimensional Data Assimilation in a limited area model. Part I: Experiments with synoptic-scale data. *Mon. Wea. Rev.*, **118**, 1250-1277.

Zuo X., Y.-H. Kuo and Y.-R. Guo., 1995: Assimilation of atmospheric radio refractivity using a nonhydrostatic adjoint model. *Mon. Wea. Rev*, **123**, 2229-2249.

COST Action 712: *Microwave Radiometry*

CHRISTIAN MÄTZLER
University of Bern, Switzerland.

Key words: Microwaves, radiometer, passive remote sensing, radiative transfer, retrieval, atmosphere, meteorology, COST

Abstract: COST Action 712 is an example of fruitful collaboration in Europe, here on scientific and technical issues of passive remote sensing at microwave frequencies for applications in meteorology. The topic is introduced, main achievements are summarised, and examples are presented for illustration.

1. WHAT IS COST?

1.1 General

COST stands for **Co**-operation in **S**cientific and **T**echnical Research (http://cost.cordis.lu/src/home.cfm). Founded in 1971, COST is an intergovernmental framework for scientific and technical co-operation, allowing European co-ordination of nationally funded research. COST Actions (the research units of COST) consist of basic, non-competitive research as well as activities of public utility. The goal of COST is to ensure that Europe holds a strong position in scientific and technical research for peaceful purposes, by increasing European co-operation and interaction, allowing access for institutions from non-member countries.

Today nearly 200 actions involve some 40000 scientists from 32 member countries and from nearly 50 institutions from additional 14 countries.

To emphasise that the initiative comes from the scientists and technical experts themselves and from those with a direct interest in furthering international collaboration, COST actions are launched on a "bottom-up" approach. Members participate on a flexible "à la carte" principle.

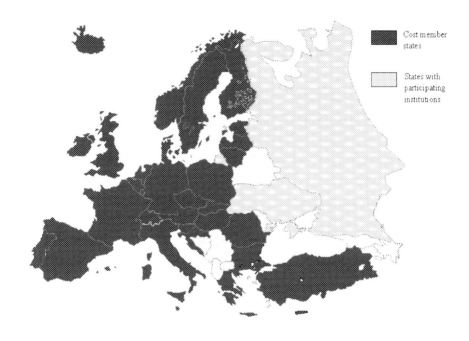

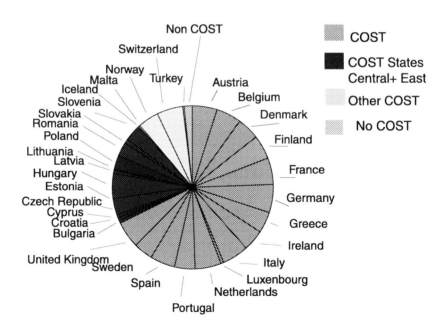

Figure 1: Map of COST Countries in Europe.

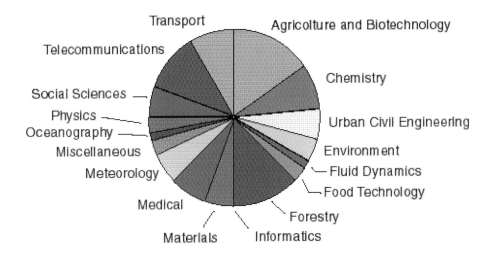

Figure 2: National participation in actions (top), and the domains of COST (bottom).

1.2 COST organisation and structure

The main bodies of COST are:

Committee of Scientific Officials (CSO) The CSO is the decision making and highest body in COST. It is composed of COST member-state representatives.

Technical Committees (TC) are responsible for a particular domain under the authority of the CSO and has mainly to prepare proposals for research projects. The TC are also responsible for technical preparation and implementation of the projects as well in an advisory capacity in the co-ordination and evaluation of ongoing projects.

Management Committees (MC) are in charge of the implementation, supervision and co-ordination of the actions. A MC is composed of one or two representatives of each signatory country.

COST National Coordinator (CNC)
One member of the CSO from each member state acts as national coordinator (CNC) for COST projects. The CNC provides the liaison between the scientists and institutions in his country and the Council COST secretariat.

Secretariat
The secretariat is provided by the Secretariat General to the Council of the European Union (secretariat of the CSO) and of the European Commission (scientific and administration matters).

2. PAST AND CURRENT ACTIONS IN THE DOMAIN METEOROLOGY

Table 1: *Past* and current actions, status November 2001.

No.	Action Title
70	*European centre for medium-range weather forecast (ECMWF)*
72	*Measurement of precipitation by radar*
73	*Weather radar networking*
74	*Utilisation of VHF/UHF radar wind profiler networks for improving weather forecasting in Europe*
75	*Advanced Weather Radar Systems*
76	*Development of VHF/UHF Windprofilers and Vertical Sounders for Use in European Observing Systems*
77	*Application of remote sensing in agrometeorology*
78	*Development of nowcasting techniques*
79	*Integration of data and methods in agroclimatology*
710	*Harmonisation in the pre-processing of meteorological data for dispersion models*
711	*Operational applications of meteorology to agriculture, including horticulture*
712	*Microwave Radiometry*
713	*UV-B Forecasting*
714	*Measurement and use of directional spectra of ocean waves*
715	Urban meteorology applied to air pollution problems
716	Exploitation of ground-based GPS for climate and numerical weather prediction applications
717	Use of radar observation in hydrological and NWP models
718	Meteorological applications for agriculture
719	The use of geographic information systems in climatology and meteorology
720	Integrated ground-based remote sensing stations for atmospheric profiling
722	Short range forecasting methods of fog, visibility and low clouds

3. COST ACTION 712: MICROWAVE RADIOMETRY – APPLICATION TO ATMOSPHERIC RESEARCH AND MONITORING

3.1 Microwave radiometry

Radiometer and brightness temperature

A radiometer is an instrument to measure the power of thermal radiation in a given frequency interval $(f, f+\Delta f)$ where Δf is the bandwidth of the radiometer. The brightness temperature T_b is the equilibrium temperature T measured by the thermometer for a perfect system (no loss, perfectly

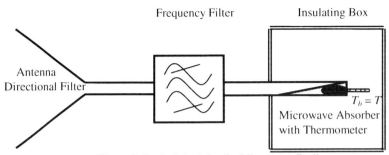

Figure 3: Basic Principle of a Microwave Radiometer

insulating box). Due to the low power ($P=kT\Delta f$, k = Boltzmann Constant) received by the antenna, electronic amplification and detection are usually required in radiometers. Thermal radiation is expressed by the brightness temperature (equivalent blackbody temperature) T_b. The sensitivity (random error) ΔT_b of the radiometer is given by the radiometer formula

$$\Delta T_b = \frac{T_b + T_{sys}}{\sqrt{\tau \cdot \Delta f}}$$

where T_{sys} (typically 100 to 1000 K) is the system noise temperature and τ is the integration time. As a typical example we assume $T_b = T_{sys} = 300\text{K}$, $\tau = 1\text{s}$ and $\Delta f = 100\text{MHz}$; then $\Delta T_b = 0.06\text{K}$.

Kirchhoff's Law and radiative transfer

Kirchhoff's Law on thermal radiation is the key to understand the information content of the observable T_b in Local Thermodynamic Equilibrium (LTE, being applicable in most parts of the terrestrial system). In the general form described by Planck ([11], § 43-46) the observable space is divided into N volume elements V_i of constant physical temperature T_i. Kirchhoff's Law [11,12] states that

$$T_b = \sum_{i=1}^{N} a_i T_i; \quad \text{where} \quad \sum_{i=1}^{N} a_i = 1$$

Here a_i is the absorptivity, i.e. the fraction of absorbed power in V_i of the transmitted power if, in the reciprocal situation, the radiometer is regarded as a stationary transmitter with the same antenna, frequency, bandwidth and position as the radiometer. Often, a_i is also called emissivity e_i of Volume V_i. In radiative-transfer theory [3,13,14] the sum over N regions is extended to an integral equation using infinitesimal steps. An example with an isothermal atmosphere is shown in Figure 4 ($N=4$). Thus, microwave radiometry requires knowledge on microwave absorption and propagation (including

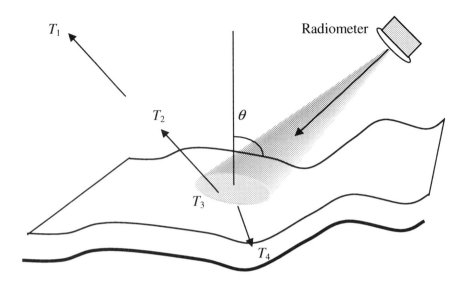

Figure 4: Geometry of microwave radiometry of the earth surface with $N=4$: T_1 is the cosmic background temperature ($\cong 3$ K), T_2 the temperature of the atmosphere, T_3 and T_4 are surface and subsurface temperature, respectively.

scattering and transmission) and on the thermal state of the medium. The potential of microwave radiometry is to gain geophysical information (T_i and/or a_i; $i=1$ to N) from measurements of T_b, using multiple frequency, bandwidth, beamwidth, polarisation, incidence angle, and observation time. The task is to establish the optimum system parameters for given applications and to define the retrieval algorithms, including the error budget.

3.2 Signatories

The following 7 countries participated in COST Action 712 on microwave radiometry (1996-2000): Finland, France, Germany, Italy, Switzerland, The Netherlands, UK.

3.3 Objective and Project Definition

The objective of COST 712 was to study and improve meteorological applications of microwave radiometry, especially with respect to the hydrological cycle, and atmospheric processes involved in these phenomena [1].

Achievement of the objective was sought through developments in four research areas, and thus in four projects:

1. The interaction of microwave radiation with the Earth's atmosphere and surface was studied in **Project 1**, *Development of Radiative Transfer Models*.
2. The retrieval, analysis and assimilation techniques through which atmospheric and surface parameters are estimated from the data was treated in **Project 2**, *Retrieval, Analysis and Assimilation Techniques*.
3. Verification and validation studies through which the accuracy and characteristics of the data analysis techniques may be assessed, were done in **Project 3**, *Assessment of Methods*.
4. Recommendations for measurement facilities and techniques, including ground-based, airborne and space-based radiometers were obtained in **Project 4**, *Measurement Facilities and Techniques*.

   Kick-off workshops were held to set up project groups, to exchange information about ongoing work and the state of the art, to define out standing items, priorities, studies needed to solve these items, and work recommended to the partners of COST 712.

3.4 Situation at the Start of the Action

When COST Action 712 started in 1996, microwave radiometers had been flown on satellites for some 25 years already. Sensors operating included the Microwave Sounding Unit (MSU) on the NOAA satellites, the Special Sensor Microwave Imager (SSM/I) on the DMSP satellites and the microwave radiometer of ERS 1. Also, data from new instruments, such as the Advanced Microwave Sounding Unit (AMSU), were known to become available soon. Research activities to use these data had been undertaken in Europe, leading to fruitful results, and some applications were already operational. However, the potential of these data was far from being fully exploited, and the interaction between users and researchers in Europe was far from being optimal. Extensive efforts were required to develop powerful data analysis procedures. For these activities support was sought through COST to coordinate research in Europe, to strengthen, to link and to focus national efforts. Although COST 712 was a small action of only 7 participating countries, for the first time, experts of remote sensing and meteorology joined to advance microwave radiometry on a European level.

3.5 Project 1

The first workshop, held for Project 1, took place at EUMETSAT, Darmstadt (D) in April 1997. Following key-note presentations on spectroscopic databases, radiative transfer and fast models in the clear atmosphere, in clouds and for surface emission, outstanding items and priorities for recommended work were discussed within subgroups (a) for the clear atmosphere, (b) for clouds and precipitation and (c) for the surface. The results are contained in the first report of Project 1 [2]. Work evolved mainly in the subgroups until their progress was presented at a follow-on meeting in Spiez (CH) in January 1999. Significant progress was achieved in all types of models, but incomplete information was identified mainly in the assessment of the accuracy.

Clear atmosphere

Results of the clear atmosphere included an improved description of the continuum absorption of water vapour in the atmosphere, new spectroscopic data in the mm- and submm- wavelength range, and software for rapid spectroscopic data simulation using existing data bases. In addition the status and contents of the relevant data bases were reviewed and recommendations were given (Table 2).

Table 2: Recommended models of the clear atmosphere

HITRAN	http://www.HITRAN.com
GEISA	http://www.ara.polytechnique.fr http://ara01.polytechnique.fr/registration
MASTER-SOPRANO	ftp from LPPM (Paris-Orsay), address: 193.55.16.151, user: anonymous, password: any, cd pub/ESA
JPL Line Database	http://spec.jpl.nasa.gov
SAO Line Database	The Smithsonian Astrophysical Observatory (SAO) database: http://firs-www.harvard.edu/www/sao92.html
MIT Transmittance Models	Fortran Codes by anonymous ftp at: mesa.mit.edu dir: phil/lbl_rt (line-by-line absorption) phil/rapid_rt (rapid transmittance)
BEAM	Fast code, combining HITRAN and JPL, Feist and Kämpfer [18], to request BEAM: http:/www.unibe.ch/~feist or send email to dietrich.feist@mw.iap.unibe.ch.
RTTOV-5 fast code	To request RTTOV-5: send email to data.services@ecmwf.int.

Clouds and precipitation

For the cloudy atmosphere, results showed that one-dimensional radiative transfer models with known profiles of scattering and emission properties are quite mature. Although three-dimensional effects were shown to be important, their impact on the remote sensing of precipitation is less significant than improved information about hydrometeor size distribution, profiles and scattering properties. Uncertainties still persist with regard to the scattering and emission from mixed ice/water phases, mainly because the knowledge on the shape and type of mixtures is incomplete. Improved

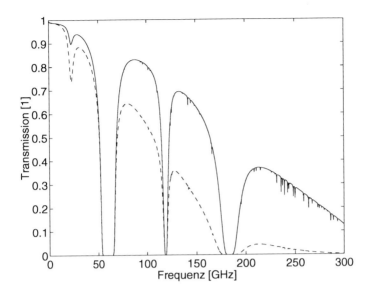

Figure 5: Transmissivity spectrum from 0 to 300 GHz through the clear mid latitude mean summer (dashed) and winter (solid) atmosphere, computed with BEAM [3,18].

physical cloud models are needed. Examples from a new scattering database showed polarisation signatures of simple spherical and ellipsoidal rain drops.

More work is required to set up a comprehensive scattering database for hydrometeors.

The following conclusions were found:
- Ongoing research in cloud and precipitation makes it difficult to recommend specific radiative transfer models
- Development of scattering data base for non-spherical particles is in progress
- The proper selection of hydrometeor profiles (and thus of cloud physics) has largest effect on brightness temperatures
- 3-D effects are noticeable, if neglected they lead to underestimates of rain rate
- Further work should concentrate on improved process models related to the formation of clouds and precipitation.

Information on the state and availability of the scattering data base: Meteorol. Institute, Universität Bonn, Auf dem Huegel 20, D-53121 Bonn, csimmer@uni-bonn.de

Surface emission

Downward looking radiometers must distinguish between contributions from the atmosphere and surface. Thermal microwave emission of the sur

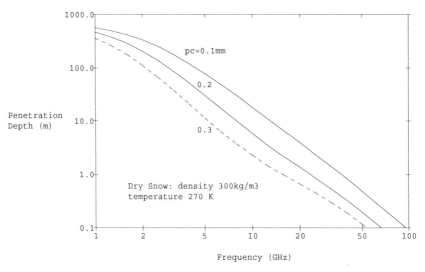

Figure 6: Illustration of the scene reflectivity r versus thickness of the surface cover for two different values of absorption coefficient of the surface layer in the zero-order (dashed) and multiple-scattering model (solid line); soil reflectivity = 0.3, scattering coefficient =1/m

face is determined by the surface emissivity and by the temperature of the relevant surface layer whose thickness corresponds to the penetration depth. Spectra of penetration depths are shown for snowpacks (Figure 7). Due to the different character of terrestrial surfaces (ocean, ice, snow, desert, forest, grass, agricultural, urban) a large number of models was needed. Depending on the application, a range of models is required with varying degree of accuracy. Figure 6 shows a comparison of two commonly used reflectivity models of scattering and absorbing surface layers.

Surface emission of ocean and land was studied, including improved dielectric data for ocean water and for land-surface components.

Roughness and relief effects were assessed. For the description of surface roughness, a satisfactory formalism (simple and sufficiently accurate) is still missing for microwave radiometry.

The often used Ruze Formula [16] of antenna theory

$$r = r_F \exp\left[-(2k\sigma_h \cos\theta)^2\right]$$

is physically incorrect for the reflectivity r of a rough surface in radiometry because it describes coherent reflection only. Here r_F is the Fresnel Reflectivity, k the wave number, σ_h the surface-height standard deviation and θ the incidence angle. For r_F=1 there is no absorption, and thus we should find r=1, independently of k, σ_h and θ. The above equation does not fulfil this condition. To avoid the problem, empirical treatments have been used. A physical solution was recently found for rough landscapes with

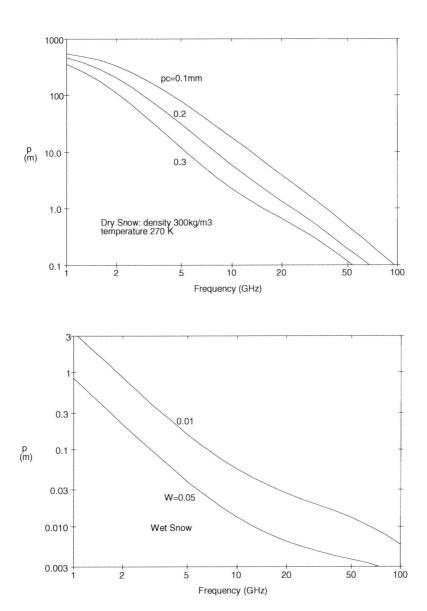

Figure 7: Penetration depth d_p in snow (density 300 kg/m^3) versus frequency as computed from the Microwave Emission Model of Layered Snowpacks (MEMLS, [15]);
a) dry snow; the grain-size dependence is expressed by the correlation length p_c, where $p_c \cong 0.1$mm is typical for fine-grained snow and $p_c \cong 0.3$mm for coarse-grained snow, such as depth hoar.
b) wet snow ($p_c = 0.3$mm) for liquid water content of 1% (solid) and 5% (dashed) by volume.

Lambertian surface facets of reflectivity r_L [17]. The reflectivity r of the landscape can then be expressed by

$$r = r_L \frac{\overline{1-\cos^2 \theta_H}}{\overline{1-r_L \cos^2 \theta_H}}$$

where θ_H is the zenith angle of the horizon (Figure 8), and the overbar means spatial and azimuthal averaging. Again, for $r_L = 1$, there is no absorption, and thus $r=1$. Figure 9 shows that this behaviour is well described by the new model. For a flat surface ($\cos\theta_H=0$) there is no relief effect, thus $r=r_L$.

Project 1 also considered effects due to surface temperature and surface covers (snow, sea ice, soil and vegetation). Still in a development phase is the optimum characterisation of the complex shape and physical properties of the surface layer, both on sea and on land. Dielectric and scattering properties strongly depend on the structure of this layer, and the same is true for the physical interactions taking place. Therefore consistent characterisations for the description of surface-atmosphere interactions and radiative transfer are needed.

The search for empirical emissivity data led to a catalogue of emissivities from which fast models and statistical data on emissivities of various surface types can be derived. The results are contained in [3].

3.6 Projects 2 and 3

The projects on retrieval, analysis, assimilation and assessment methods worked jointly, starting their activities with a workshop held at ESTEC, Noordwijk (NL) in February 1998. Key-note presentations included the retrieval of geophysical information from imaging and sounding radiometers, the assimilation of such data in numerical weather prediction models, model validation and assessment of the methods. The issues were discussed within the two groups (Projects 2 and 3), leading to recommendations for further improvements.

Figure 8: Landscape, consisting of Lambertian surface facets, with horizon at zenith angle θ_H.

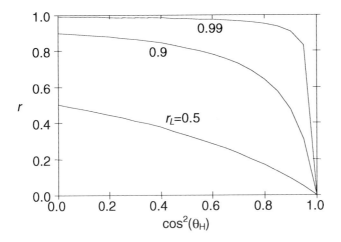

Figure 9: Landscape reflectivity r versus average $\cos^2\theta_H$ for 3 different values of r_L: 0.99 (top), 0.9 (middle) and 0.5 (bottom curve).

The results of this workshop were compiled in an ESA Workshop Publication [4]. A recommendation was to extend the methods of Optimal Estimation or Bayesian Estimation to all types of retrievals. Combination of radiometer data with other types of information (e.g. from radar or visible and infrared radiometers in case of rain) was recommended to help reduce ambiguities, especially when many parameters are involved. The assimilation of improved process models within the data analysis is needed to further reduce ambiguities and retrieval errors. These findings further emphasise the need for accurate forward models.

Methods to rigorously characterise systematic and random errors are to be developed. This is especially important in view of the increasing trend toward direct assimilation of remote sensing data into numerical models. Since errors also depend on the stability and accuracy of the instruments, the importance of state-of-the art calibration references was mentioned, too.

Much of the required techniques for retrieval, analysis and data assimilation were already mature at the time of the starting workshop of Projects 2 and 3. An excellent textbook on retrieval techniques was published during the action [5]. Therefore the work of Projects 2 and 3 did not have to concentrate on the methodologies, but rather on results to be expected with the advent of new sensors. Indeed, the impact of the temperature sounding channels of AMSU-A for numerical weather prediction was striking, not only in the southern, even in the northern hemisphere. Progress - such as this one - was presented at the second workshop of Projects 2 and 3, held in Amsterdam in January 2000, dedicated to new sensors, new applications and new retrievals. The results are contained in the second report of Projects 2 and 3 [6].

After discussing the choice between "retrieved products" and "radiance" assimilation, the report is composed of two parts: the first one is devoted to new sensors, the Soil Moisture and Ocean Salinity (SMOS) mission, and fully polarimetric radiometers. The second part is concerned with progress using current radiometers: the Tropical Rainfall Measuring Mission (TRMM, http://kids.earth.nasa.gov/trmm/) has improved the understanding of the retrieval of precipitation information. Interest is growing in the assimilation of such information in meteorological models. Another development is related to retrieving "standard" atmospheric quantities, but over non-water surfaces: it is shown that possibilities exist over polar ice. It is also shown that surface properties can be obtained from SSM/I: latent heat flux over ocean with a better accuracy than previous estimates, and land surface emissivity, which opens the possibility to retrieve atmospheric contents. Profiling of atmospheric temperature with radiometers operating in the absorption band of oxygen near 60 GHz were shown to be very successful, especially with the AMSU-A. Effects of clouds need to be taken into account. Higher-frequency channels could possibly give information on ice-cloud properties, primarily the ice-water path.

Further achievements on microwave radiometry were presented at the 6th specialist meeting on microwave radiometry at Florence (I) in March 1999. Reviewed papers of this conference were published in [7].

3.7 Project 4

Project 4 started by collecting information on existing facilities for microwave radiometry. The list is being updated and presented by the UK Met Office (http://www.met-office.gov.uk/sec5/COST712/survey.html). The work of Project 4, consisting of two workshops, was focused on spectral ranges where efforts were most urgently needed, namely the protected L-Band channel at 1.41 GHz for soil-moisture and ocean-salinity measurements, and frequencies above 100 GHz, mainly for retrieving atmospheric information from trace gases and ice clouds.

Radiometry in the mm- and submm- wavelength range

The first workshop, held at the International Space Science Institute in Bern (CH) on Dec. 9-10, 1999, was dedicated to the techniques in the mm- and submm- wavelength range. Experts discussed critical aspects, covering antenna design and measurements, quasi optics for wave guiding, low-noise mixers, and spectrometers for high-quality analysis. Problem areas were identified. Today the bandwidth of radiometers and spectrometers is limited by mixer and spectrometer performances to 1 or a few GHz. Beyond this range hybrid systems with multiple receivers are used. Very broadband instruments to be developed would allow the simultaneous observation of many trace gases. Key problems identified were inadequate calibration loads, baseline issues and the optimisation and determination of the beam efficiency of large antennas. Some of the problems are linked, e.g. poor calibration loads will lead to internal reflections which in turn will produce frequency dependent standing waves and thus baseline effects. Good calibration loads were considered as being of utmost importance.

The results of the workshop form the first part of the Project 4 Report [8].

L-Band radiometry

It is now well accepted that L-Band (frequency range: 1 to 2 GHz) radiometry is the most appropriate way to remotely estimate soil moisture and sea-surface salinity. At the meeting at ESA ESTEC, Noordwijk, March 6-7, 2000, three areas were considered, namely space missions, ground-based and airborne L-Band radiometers for validation purposes, and suitable test sites for measurement campaigns. The space missions included plans of the United States (ESTAR, HYDROSTAR) and of ESA (SMOS), both using interferometric techniques to achieve the required spatial resolution (for the scientific part of SMOS, see [6]). The present challenge is to realise such techniques with the required accuracy and stability. Another concept relying on the deployment in space of a very large antenna is studied by the NASA Instrument Incubator Program. A review of existing radiometers (ground-based and airborne) and of facilities for testing and for campaigns was carried out after the workshop. The resulting information forms the second part of the Project 4 Report [9].

3.8 Achievements and future needs

Achievements of COST Action 712 were presented and discussed - from the remote sensing and application points of view - at the Final Symposium held at the European Centre for Medium Range Weather Forecasts (ECMWF) in Reading (UK) on November 21-22, 2000 [10]. Possible topics for new actions were identified: Remote sensing for hydrology, polarimetric techniques in microwave radiometry, and stratosphere-troposphere interactions. Furthermore it was stated that activities must continue, especially on the protection of frequency bands from man-made noise, the development, validation and optimisation of radiative transfer models and of process models, and the development and maintenance of spectral databases.

4. REFERENCES

Chandrasekhar S., Radiative Transfer, Dover Publication (1960).
English S., A report on the COST-712 Final Symposium, FSreport.doc, Nov. (2000).
Eymard L. and C. Mätzler (eds.), Toward new sensors and new retrievals in radiometry, Report following the COST712 Project 2+3 Workshop in Amsterdam, 18-19 January 2000, Nov. (2000).
Feist D. G. and N. Kämpfer, BEAM: a fast versatile model for atmospheric absorption coefficients from 0-1000 GHz, In Tadahiro Hayasaka, Dong Liang Wu, Ya-Qiu Jin, and Jing shan Jiang (eds.), Microwave Remote Sensing of the Atmosphere and Environment, Proceedings of SPIE, The International Society for Optical Engineering, Vol. 3503, pp. 301-312 (1998).
Janssen M.A. (ed.), Atmospheric remote sensing by microwave radiometry, John Wiley, New York (1993).
Kämpfer N. (ed.), Measurement facilities and techniques in the millimeter and submillimeter wavelength range, COST Action 712, Report of Project 4, Part 1, Nov. (2000).
Kerr Y. and M. Hallikainen (eds.), L-Band radiometry – Measurement facilities and techniques at low frequency, COST Action 712, Report of Project 4, Part 2, Nov. (2000).
Kirchhoff G., Über das Verhältnis zwischen dem Emissionsvermögen und dem Absorptionsvermögen der Körper für Wärme und Licht, Poggendorfs Annalen der Physik und Chemie, Bd. 109, S. 275-301 (1860).
Mätzler C. (ed.) Development of radiative transfer models, COST Action 712, Report from Review Workshop 1, EUMETSAT, Darmstadt, Germany, April 8 to 10, 1997, Oct. (1997).
Mätzler C. (ed.) Radiative transfer models for microwave radiometry, COST Action 712 Application of microwave radiometry to atmospheric research and monitoring, Meteorology, Final Report of Project 1, European Commission, Directorate General for Research, EUR 19543, ISBN 92-828-9842-3 (2000).
Mätzler C. and A. Standley, Relief effects for passive microwave remote sensing, Internat. J. of Remote Sensing, Vol. 21, No. 12, pp. 2403-2412 (2000).
Memorandum of Understanding for the Implementation of a European Concerted Research Action Designated as COST Action 712 Microwave Radiometry, COST 281/95, Brussels (1995).
Pampaloni P. and S. Paloscia (Eds.), Microwave Radiometry and Remote Sensing of the Earth's Surface and Atmosphere, Proc. 6th specialist meeting, Florence, Italy, March 16-18, 1999, VSP, Zeist, The Netherlands (2000).
Planck M., Theorie der Wärmestrahlung, Vorlesungen, 6. Aufl., Leipzig (1966).
Rodgers C.D., Inverse methods for atmospheric sounding, F.W. Taylor (ed.) Series of Atmospheric, Oceanic and Planetary Physics - Vol. 2, World Scientific Publ. Co., Singapore (2000).
Ruze J., Antenna-Tolerance Theory – a Review, Proc. IEEE, 54, 633-640 (1966).
van Oevelen P.J., L. Eymard and S. English (eds.) Microwave radiometry for atmospheric research and monitoring, COST 712 Workshop, ESTEC Noordwijk, The Netherlands, 9-11 Feb. 1998, ESA-Workshop Publication WPP-139, May (1998).
Wiesmann A., and C. Mätzler, Microwave emission model of layered snowpacks, Remote Sensing of Environment, Vol. 70, No. 3, pp. 307-316 (1999).

Advances in Global Change Research

1. P. Martens and J. Rotmans (eds.): *Climate Change: An Integrated Perspective.* 1999
   ISBN 0-7923-5996-8
2. A. Gillespie and W.C.G. Burns (eds.): *Climate Change in the South Pacific: Impacts and Responses in Australia, New Zealand, and Small Island States.* 2000
   ISBN 0-7923-6077-X
3. J.L. Innes, M. Beniston and M.M. Verstraete (eds.): *Biomass Burning and Its Inter-Relationships with the Climate Systems.* 2000   ISBN 0-7923-6107-5
4. M.M. Verstraete, M. Menenti and J. Peltoniemi (eds.): *Observing Land from Space: Science, Customers and Technology.* 2000   ISBN 0-7923-6503-8
5. T. Skodvin: *Structure and Agent in the Scientific Diplomacy of Climate Change.* An Empirical Case Study of Science-Policy Interaction in the Intergovernmental Panel on Climate Change. 2000   ISBN 0-7923-6637-9
6. S. McLaren and D. Kniveton: *Linking Climate Change to Land Surface Change.* 2000
   ISBN 0-7923-6638-7
7. M. Beniston and M.M. Verstraete (eds.): *Remote Sensing and Climate Modeling: Synergies and Limitations.* 2001   ISBN 0-7923-6801-0
8. E. Jochem, J. Sathaye and D. Bouille (eds.): *Society, Behaviour, and Climate Change Mitigation.* 2000   ISBN 0-7923-6802-9
9. G. Visconti, M. Beniston, E.D. Iannorelli and D. Barba (eds.): *Global Change and Protected Areas.* 2001   ISBN 0-7923-6818-1
10. M. Beniston (ed.): *Climatic Change: Implications for the Hydrological Cycle and for Water Management.* 2002   ISBN 1-4020-0444-3
11. N.H. Ravindranath and J.A. Sathaye: *Climatic Change and Developing Countries.* 2002   ISBN 1-4020-0104-5; Pb 1-4020-0771-X
12. E.O. Odada and D.O. Olaga: *The East African Great Lakes: Limnology, Palaeolimnology and Biodiversity.* 2002   ISBN 1-4020-0772-8
13. F.S. Marzano and G. Visconti: *Remote Sensing of Atmosphere and Ocean from Space: Models, Instruments and Techniques.* 2002   ISBN 1-4020-0943-7

KLUWER ACADEMIC PUBLISHERS – DORDRECHT / BOSTON / LONDON